*Technological Innovation
and Society*

Technological Innovation and Society

Edited for the Columbia University Seminar

On Technology and Social Change by

Dean Morse and Aaron W. Warner

 *1966*

Columbia University Press NEW YORK AND LONDON

Preface

THE PAPERS AND DISCUSSIONS in this book represent the third year of deliberations of the Columbia University Seminar on Technology and Social Change, which was formed in 1962. Although subjected to a fair amount of editing, this material continues to reflect the informal but incisive quality, and the spirited give and take, which characterize the seminar proceedings.

The reader may wonder about the nature of the seminar. It brings together for regular monthly meetings throughout the academic year a diverse group of physical scientists, social scientists, business leaders, and public officials, and is part of the University Seminar movement at Columbia University that first began in 1944 under the leadership of Professor Frank Tannenbaum, who has since guided the movement through a period of rapid and significant growth. There are now thirty-nine seminars with a total of over eleven hundred members. Each University Seminar, an independent "intellectual community of fellows," has received official sanction and permanent status from the University. Each seminar devotes itself to the study of a particular subject that is of interest to its own group.

The Columbia University Seminar on Technology and Social Change follows a course that is mapped out by its steering committee. During the first year the members sifted out the major areas of concern and suggested lines for future inquiry. The proceedings of that year are contained in *Technology and Social Change*, edited by Eli Ginzberg. In its second year the seminar undertook to analyze and unravel some of the complexities of the interrelationship between science and technology. The results are recorded in *The Impact of Science on Technology*, edited by Aaron W. Warner, Dean Morse, and Alfred S. Eichner. In its third year the seminar has attempted to relate technology to innovation and to begin an exploration of the impact of technology on specific aspects of the social environment. The range of issues taken up in a given year is obviously limited by the number of meetings, and this volume is, therefore, only one part of a continuing exploration of the frontiers of change in a world in which technology plays an increasingly dominant role.

As editors we express our profound gratitude to the speakers and the members of the seminar who have so generously assisted us in preparing their manuscripts and revising the texts for publication. We also express our appreciation for the arrangements made by the U.S. Department of Labor, Office of Automation, Manpower, and Training, to disseminate widely copies of this and of previous books, thereby assuring that the discussions reach a larger audience of individual persons and groups who have a vital interest in the issues explored.

DEAN MORSE

AARON W. WARNER

Columbia University
February, 1966

Contents

PART ONE

Science and the Process of Innovation

Introduction

THE AUTHORS of the four papers in Part One, all scientists, share a concern with the interrelationship between science and technology; a concern that has manifested itself in their active work as scientists. E. R. Piore and Robert L. Hershey have both had the general responsibility of overseeing the scientific and research activities of very large corporations, which have played leading roles in the translation of sophisticated scientific research into a practical and highly successful technology. Edwin Land, the founder and president of the Polaroid Corporation, has with conspicuous success been able to initiate scientific research that has eventuated in strikingly original and popular commercial products. Moreover, Mr. Land, combining in his own person the role of scientist and inventor, has given to the Polaroid Corporation a unique character in which a high degree of sensitiveness to the creative potential of human beings is blended with a keen awareness of the technological possibilities that are the fruit of scientific discovery. Jerome B. Wiesner, Dean of Science at the Massachusetts Institute of Technology, was President Kennedy's scientific adviser at a time when there was an intense and growing national concern with the relationship between science, technology, and public policy.

Having explored various aspects of the relationship between

science and technology in the previous year, the Columbia University Seminar on Technology and Social Change turned, in 1964-1965, to this group of scientists for their reflections on the interrelationship between science and technology, on the one hand, and for their conceptions of the problems associated with the application of scientific and technological developments in the world of business and commerce on the other hand. Before turning its attention to some of the more general questions created by the impact of technology upon society, the seminar felt it was appropriate to get a general sense of how scientific discovery becomes actual technological advance in the economy.

Indeed, the title of Mr. Wiesner's post in the executive branch of the federal government, Scientific Adviser to the President, is an acknowledgment that the relationship between the development of science and other elements of society—particularly the economic and the military, but by no means confined to these—is today of special importance, and raises new and difficult questions involving public and private policy of very diverse character. The post is at the same time an expression of several fundamental preconceptions of our time.

First of all, there is a widespread belief that industrial preeminence and along with it, economic growth depend to an important degree upon the character of a nation's scientific activity. This belief is not a recent development. It was, for example, very strongly expressed in Alfred Marshall's *Industry and Trade* some fifty years ago, and at the turn of the century it was often brought up in discussions centering on England's gradual loss of industrial leadership to Germany and the United States. German and American scientific preeminence, according to this view, was being rapidly and inevitably translated into technological and industrial leadership; the explanation of England's decline was the failure of her educational system to provide an adequate supply of scientists, and was coupled with the lack of appreciation by England's business leaders of the importance of science.

But if the belief that scientific leadership is essential to industrial leadership has long since been accepted, American anxiety on this score is a relatively recent development. It is not difficult to account for this heightened awareness of the relationship between science and economy. Moreover, the assertion that the growth rate of the United States economy was falling behind that of the Soviet Union, and that the Soviet Union might be in a peculiarly favorable position to apply the most advanced scientific discoveries to her economy was reinforced by the spectacular Russian successes in rockets, satellites, and atomic weapons. Forecasts of gross national product, in which the highest growth rates achieved by the Soviet Union were projected into the future alongside the relatively slow growth rates of the United States during the late 1950s, gave credibility to fears that in the near future Russia might overtake the United States economically. The military and political consequences of such a shift in relative economic strength were spelled out, and popular opinion tended quite naturally to associate the predicted national peril with the possibility that American science was lagging or was misdirected in its efforts.

Finally, it may be conjectured that the very size of the contemporary American scientific establishment has heightened apprehension about the interaction between scientific activity and society in general. The extraordinary scale and complexity of modern scientific activity, which quite outstrips the capacity of what even the most informed scientific mind can comprehend in any precise fashion, has tended to reinforce an attitude which has felicitously been labeled "technophobia."[1] Believing that science is ethically neutral and has no limitations, the nonscientist is apt to agree with Kingsley Martin's statement, "If

[1] Gerald Leach, "Technophobia on the Left: Are British Intellectuals Anti-Science?" *New Statesman*, LXX (August 27, 1965), 286-87.

they have not destroyed the world first, our scientific managers may yet create a stable and terrible world in which common men play an even smaller and more servile part."[2]

Of course the technophobe is not alone in his misgivings about science. In the past few years a loosely defined group of physical scientists, social scientists—economists and sociologists for the most part—and others, including a number of theologians, have been asking whether the rate of technological change has or is about to increase so sharply as to challenge the capacity of our major social institutions to adapt to it. Is it placing upon certain groups and individuals, ill-equipped to cope with the changes that will be required in their ways of life, a disproportionate share of the social cost of technological change? The intensity of the debate on this question is evidenced by a spate of articles and books. Our evaluation of the adequacy of national policy in a number of vital areas depends upon how we interpret the evidence, indeed, it depends upon what we deem to be relevant evidence. At stake, for example, is the adequacy of fiscal and monetary policy and the appropriateness of the poverty program.

Against this background of national concern about science and technology it is of some interest to see how a few distinguished scientists (admittedly not representative in any sense, in fact unrepresentative because some of their major concerns have been just the interaction between science and society, either in very specific or in quite general terms) view the relation between science and technology and what they think the major character of the social impact of science and technology is apt to be.

At the outset it can be said that these four scientists do not express the belief that a rapid increase in scientific knowledge

[2] Kingsley Martin, "The Failure of Progress," *New Statesman*, LXX (July 2, 1965), as quoted by Leach, "Technophobia on the Left," *New Statesman*, LXX (August 27, 1965), 286.

coupled with an increase in the rate of technological change gives serious ground to fear a social disaster. Quite the contrary, the general tone of their remarks is guardedly optimistic. Technological change, it is true, may produce problems of adjustment, but in no sense should society attempt to slow down the rate of technological change in order to cope with social problems. Indeed, so strong is this positive attitude to the possibilities opened up to society by scientific progress that in the discussions that followed the presentation of the papers, several members of the seminar whose concerns center on the social impact of technological change occasionally criticized the scientist contributors, as a group, for a lack of concern for or awareness of the social impact of scientific and technological progress.

In justice to the scientists, however, it should be emphasized that they are indeed concerned, but that, in general, this concern takes a quite specific form. The scientific mind, or at least this sample of it, searches out particular points at which scientific and technological change may cause problems and then looks for particular remedies for these problems. And, as a corollary, the scientific mind seems generally to hold that a series of particular solutions to particular problems may suffice, in the long run, to bring about the necessary adjustment between technological advance and society.

Not only do the scientists represented in this book tend to have a mildly optimistic view, and a predilection for specific and short-term solutions to problems as they emerge, but also, as might be expected, they tend to eschew long-term predictions. The scientific mind, as shown in these four papers, does not seem to consider itself also a prophetic mind. Moreover, a mild skepticism about a number of widespread beliefs is clearly in evidence. For example, the two scientists most directly concerned with industrial research and development activities, Mr. Piore and Mr. Hershey, agree in doubting the value of numerous efforts to categorize research effort as "pure" or "applied." From their point of view it appears impossible to pigeon hole scientific effort so neatly. Moreover, both men emphasize that "fundamental" research can in fact take place in the most unlikely areas. Similarly Mr. Wiesner is suspicious of the label, "innovation."

Much more significant, however, is Mr. Wiesner's skepticism about the belief that technological advance is firmly tied to and derived from scientific advance. Readers of the previous volume of the seminar's proceedings may recall that Harvey Brooks, Dean of the School of Applied Science at Harvard University, emphasized that the links between science and technology in the past have frequently been quite tenuous. It is interesting to note that the Dean of Science at the Massachusetts Institute of Technology feels that the same can be said about much of modern science and technology.

The scientists who contributed papers to the seminar display a pragmatic optimism, behind which can be glimpsed a general faith that reason and a scientific approach to social problems will be able to provide viable, if imperfect, solutions. It is noteworthy, however, that Mr. Land also maintains, on the basis of his experiences in the Polaroid Corporation, an attitude toward human nature, creativity, education, and, finally, toward the character of the good society itself, which in many fundamental respects runs quite contrary to conventional views.

The four papers in Part One do share a number of preconceptions and attitudes. Nevertheless, each paper has its own special point of view and its special concern. While Mr. Land is particularly concerned about the channels modern industrial society affords for the flow of creative activity, Mr. Wiesner focuses our attention upon the difficulties involved in organizing and carrying out the process of innovation as a matter of national policy. Echoing to some extent the views of Joseph Schumpeter, Mr. Wiesner emphasizes that there is indeed a special psychological type that is largely responsible for innovation.

It is also very appropriate to the subject of Part One that Mr. Hershey should address himself primarily to the concrete experiences involving innovation and the technological change which he has experienced over a number of years as a scientist-engineer in various departments of the Du Pont Company. As Mr. Hershey sees it, perhaps the single most important lesson to be derived from these experiences is the necessity for the industrial manager in charge of technological change to involve at a very early point the widest possible spectrum of individuals who will participate, at different stages, in the introduction of a new technique or process or product.

Mr. Piore addresses himself to the question of just how a large, technologically advanced firm, using a proper balance of research and development effort, survives and grows. Rather than categorizing research effort as "pure" or "applied," Mr. Piore, like Mr. Hershey, prefers to categorize research according to the role it plays in the ongoing life of the firm.

He also discusses an issue that has been present in many of the seminar's discussions from the beginning, the relationship between the decision maker, who, frequently, not scientifically trained, must make decisions involving scientific and technological activity and his advisers, who are scientists. Mr. Piore emphasizes that it is almost impossible to isolate science and technology from anything else that is taking place in modern society, and that people who are technically trained increasingly believe that all those responsible for making decisions must be exposed to a good deal of science, but that at the same time it is important to keep in mind that when a scientist participates in policy decision-making he does not do so merely in his capacity as a scientist.

The range of questions that are touched upon in the papers and discussions of Part One is impressively wide. At the same time it is evident that the answers to a number of these pressing questions are in a highly unsatisfactory state. It is perhaps characteristic of the scientific mind that this is admitted in a

matter of fact way. Fundamentally, the scientists who contributed papers to the proceedings of the Seminar on Technology
and Social Change during 1964-1965 are asking what conditions make for scientific and technological creativity and what
contribution the scientist and technologist can make to help
solve the problems associated with scientific and technological
advance. The scientists point out, however, that an understanding of the general character of science—of the relationship
between scientific discovery, technological advance, the process
of diffusion of technology into the industrial framework, and
of the preconditions necessary for scientific creativity—is essential if solutions to grave problems of national and industrial
policy are to be found, which make full use of the potential
to improve and enrich our lives that science has. None of these
scientists doubts for a moment that science has this potential.
In the 1963-1964 proceedings of the seminar Mr. I. I. Rabi
asked, "Where do we find the really new thing, the moving
thing, the thing that will show the Glory of God and the orig-
inality of nature . . . and where do we find the imagination that
delves into the mysteries of life, into the existence of matter?"
He answered, "That imagination, I would say, will be found
only in science." In Mr. Rabi's view, science is the essentially
creative activity of our age. In the same vein, Mr. Land, emphasizing that only a creative society can be considered a good
society, calls for a profound change in our attitudes in order
to make possible a release of creative energies wherever they
may exist, and points out to his fellow scientists and industrialists that here the modern industrial corporation offers an
opportunity and a challenge that we can neglect only at our
peril.

Technology and Innovation

by JEROME B. WIESNER

Dean, School of Science,
Massachusetts Institute of Technology

WE ALL have some intuitive feelings about technological innovation, and yet, when we try to specify how we might organize or carry out the process of innovation, the creation of new devices or new industries, we become rapidly confused. The story of the blind men and the elephant can be applied to technical innovation. What each man has experienced and where he has touched the animal determine his ideas about what it is. And each man has very different ideas. First of all, there is no such thing as innovation per se, because there are many ways in which one can innovate. There are many industries and many problems, and each industry and problem is quite different. We tend to oversimplify and to think that simply applying science, somehow, automatically leads to the solution of all problems. But, of course, it is much more complicated than that.

In fact, even to equate technology and science is wrong. Science is the quest for more or less abstract knowledge, whereas technology is the application of organized knowledge to help solve problems in our society. We tend to forget that even today a great deal of technology, for example, much of mechanical and of civil engineering, may not depend on science. There are many important, creative, innovative activities which, if they draw on science, draw on it incidentally. For a long time this was true of most technology. Then, as soon as people began to understand a little about electricity and chemistry, this knowledge was exploited to make devices for special purposes, for instance, the electric motor, the telegraph, the telephone, plastics. Most of these inventions were not made by scientists (they were more interested in understanding natural phenomena), but by people who somehow learned a little bit about a specific scientific field, saw an opportunity to do something about an existing problem, and wished to exploit a new idea in order to make an invention based on the new knowledge.

We have forgotten that invention is still the basic ingredient of innovation. We still need people who have the turn of mind, the interest, the inspiration, and the creativity to be inventors. Early inventors used facts from the every day world in which they lived, and relied upon mechanical devices and processes they could observe. The modern technologist, who produces a transistor or a new plastic or a new antibiotic is no less an inventor, but he has to live in a special world and have special knowledge on which to base his creations. But the possession of a scientific background does not automatically make a person into an inventor.

What is more, the question of how we produce, how we train—create, in a sense—creative people is a subject about which a great deal has been written, but which still is not really understood. It is related, I believe, to another problem we do not understand: how human beings think. If a professor is asked what he is trying to do, he will say that he is trying to teach people to think. Then if the question is raised, "How do you teach people to think?" it will be found that he really does not know. If he has any success at all in his effort, it is because of his own example. At the heart of the problem, not only of innovation, but of scientific research too, is this dilemma: some students who are very good at problem-solving and get very good grades, turn out to be completely unoriginal when put to work on a research project, and on the other hand, many a student who is not a very good scholar and who may have to be dragged through his courses and exams, really has the creative spark and often becomes a very good research worker.

So the first thing we need if we want innovation, I think, is a person with the drive and imagination and burning desire to create, to invent. Of course, innovation takes a great many other things too. The opportunity and the need must exist, and so must the funds for this kind of creativity—this latter factor is becoming an increasing stumbling block to invention throughout the world as science and technology become more and more costly.

In most of the civilized world we have a faith, which I think is justified although it is difficult to prove statistically, that the rate of growth of our economy is in some way tied to the quantity of research and development, particularly the research, that we do. Nobody will question the fact that if we had not done a great deal of *basic* research, and thus created the scientific knowledge on which our technology is based, we would not have our growing industrial society. Nevertheless, we have not established a quantitative relationship between rates of expenditure for research and rates of economic growth. Also, when it comes to *applied* research and development, we should be convinced that the things being sought are needed before we spend very large sums of money on them. This precaution is necessary, partly because applied research and development is a lot more expensive than basic research, and partly because new knowledge, fundamental understanding, is an indestructible asset, which can be used at any time in the future. Whether it is acquired today, or five or ten years from now is not of much consequence. But the value of technological developments has a strong time factor.

Is there any clear-cut connection between the rate of research and development expenditures and the rate of economic growth? Can one place a dollar value on research results? If we think about this question for a little while, we see that it is at least partly a foolish question, because we cannot put a price on the value of something like penicillin. But there are some things that can be priced. It may be possible, for example, to establish the value of some specific technological developments, such as civilian atomic energy. But even here there are many pitfalls. How does one discount the money spent today against the year 2000? Is the population of the United States in 2000 going to be 1,000 million or half a billion? Such questions obviously have a major effect on the results of this kind of economic analysis.

There is a general belief in the United States, which I share, that as we increase industrial productivity and have more manpower available, one of the things we must do is create new products and new industries. Furthermore, we want old industries to become more efficient to help raise the standard of living. When we try to understand why some industries are very dynamic and many other are not, we find that many of the less dynamic industries have not used technology to any real extent, either to improve their productivity and their products or to create new products.

It is hard to understand why this is so. But those industries that have not been very aggressive in the application of technology and science appear to have some things in common. One is that they tend to be industries made up of many small units. The housing industry, for example, is made up of a very large number of small builders. As a result, there is no single large company which, if it invests substantially in research and development, will receive a substantial return on its investment. Even most of the manufacturers of building materials are not large enough to support effective research and development efforts. Such industries also have no traditional links with research and do not appreciate its potential. Many, such as the railroads, are in such financial difficulty that they cannot make long-term research and development investments.

In the last few years we have puzzled a good deal about whether there was a role for the federal government in fostering innovation. We examined what other countries had done in this field. We examined governmental attempts to stimulate industrial innovation in England and the Soviet Union, the only countries in which the governments have made major efforts. In my judgment both have failed.

In England, after World War II, the government established a number of industrial institutes, nearly fifty of them, whose prime purpose was to help create new products and to introduce modern technology and modern processes in British industry. This was done because the British recognized that their industry was not on a par with most modern American industry. With a great deal of imagination and energy they set out to do something about the situation. But the results of these activities in 1961 and 1962 showed that most of them had been failures. They had done a number of very interesting but largely unimportant things. This whole assembly of institutes, which had spent a great deal of money, had not developed any really important new products or new ideas that had been introduced in industry. To say that these institutes had made no contribution at all would be unfair, but it is a fact that they did not revitalize British industry, as had been the initial hope.

One reason for this unsatisfactory result is, it seems to me, plain. Having a new idea and demonstrating its feasibility is the easiest part of introducing a new product. Designing a satisfactory product, getting it into production, and building a market for it are much more difficult problems. Persuading people to take expensive gambles on new ideas can be very difficult, particularly in a situation in which the gambler does not acquire a proprietary interest in the new product. This was one of the most serious weaknesses of the industrial institutes.

If they developed an attractive product, obviously marketable, it was equally accessible to everyone. As a result, no one was prepared to take the expensive first plunge to develop the market—or to take the possible loss if it could not be developed —knowing that as soon as it was, other companies could step in.

The Soviet Union has also made very large investments in applied science laboratories, in industrial research, and in development laboratories. Yet a major preoccupation of the Soviet leadership at this time is how to make their industry more efficient and how to develop new and better products. They are not too proud to copy what they need. If they want a snow plow, they buy existing snow plows, decide which is the best one, and then copy it. But Soviet industry on the whole has not been able to copy Western products advantageously. In part, this is a result of the essential inadequacy of their industrial base, despite the great improvements made in the past decade. There is also a seeming lack of adequate interaction between Soviet scientists and technologists, and Soviet industry.

These failures are not easily understood, and I am not satisfied yet with my analysis, but I do have some further thoughts about the British case. In addition to the problem of risk-taking already discussed, the British system is not designed to attract the kind of people who are driven to make or do something new. The innovator, the man who thinks he has the great idea that will make a million dollars, is not likely to go to work in a fifty-man laboratory where the purpose is "to invent" something in the abstract.

This is a serious problem. A man cannot be ordered to have new ideas. We cannot simply say, "We want three new ideas this year and we want them in the following fields," and really expect them to be forthcoming. Innovation is a much more complex process than that. An essential missing ingredient in the British case seems to me to be the person with the inspiration to do something new. As I said earlier, the people with the inspiration are not necessarily the best trained scientists or the best trained engineers. The technical innovators are men who not only have some scientific knowledge but who are also inspired to put it to work on every new idea that comes their way. This does not mean that there are not men in most of our laboratories today who are both very good scientists or engineers, and very creative people at the same time. But most laboratory directors would admit that in order to end up with a few really creative people, they need a large staff.

In order to cultivate people who have both a scientific or technical competence and who are also capable of some inspiration, we might experiment with technological institutes spanning universities, government institutions, and industry, and which should also have sufficient funds to support the development of inventions. Also, in those industries where the main difficulty is the fragmentation into a large number of small firms, we might offer financial incentives in order to bring companies together for cooperative research and for development activities. I hope that with a bit of exploratory work on our part, and an increased understanding on the part of the Congress, those people in the government who are pressing for these experiments will be allowed to start some of them. Then we will have a chance to learn whether it is possible to stimulate innovation and technological development in a laggard industry.

DISCUSSION

SCIENTIFIC ACTIVITY AND TECHNOLOGICAL CHANGE

Question: There are people who feel they want to innovate and invent, and there are people who are engaged in scientific activity. To what extent are these people different people and to what extent are they the same people in different circumstances? I would particularly like to explore the question of what happens to the scientist in wartime, when objectives are clear, and people who normally engage in pure science and in the pursuit of knowledge apply themselves to technological problems. Is the lack of inventive creativity in scientists due to the fact that they do not see clear objectives, or is it primarily due to the fact that they are not the people who are interested in achieving objectives of that sort?

Mr. Wiesner: Many people who are good scientists could be good technologists. As a matter of fact, a great deal of science is inherently innovating, whatever else the scientist is trying to do. The scientist's work necessarily results in new experiments, new insights, new ideas. And it often turns out that when a good scientist turns his energies to technology, he becomes a better technologist than somebody who has been trained as an engineer. This was conclusively demonstrated during World War II. I think that one of the reasons for the great managerial successes of the big industrial laboratories, such as, for instance, the Bell Telephone Laboratories, is that they are able to hire scientists to do basic research and are able gradually to turn the interests of these individuals toward technology and innovation.

But the very good scientist is more likely to want to do his innovating in science. That is why there is a real social need for that strange character I spoke of—the inventor who is not terribly skilled or terribly knowledgeable, but who gets an insight based on the little scientific information he has, and who then needs a great deal of help from scientists and technologists.

Question: Supposing one could formulate social needs more explicitly, do you think that more scientists would become innovators in order to meet these social needs?

Mr. Wiesner: I think so. When you bring a group of scientists together for three months or six months or a year, as we have done from time to time when we have had a specific military problem, very substantial inventions and ideas have almost always resulted.

Question: One hears people discuss today the impact of scientific activity upon technology. What is not clear in these discussions is whether they are talking about the impact of the scientific activity of today on the technology of today, or whether what they have in mind is the impact of scientific developments today upon technology thirty years from now. If I may put it in slightly professional jargon, there may be nonlocal interactions, nonlocal in space and nonlocal in time. One person's scientific discovery is the basis for another's technological advance at a different time and place. I do not know how you would measure this kind of phenomenon, yet it is the sort of problem one constantly meets in the domain of science itself.

Mr. Wiesner: I have come to the conclusion that there is usually a period of twenty years or so between the discovery of a new scientific principle and its forceful impact on industry. In the case of the transistor, for example, it took about that long. Some things move a bit faster, but I would say that many are even slower.

For example, our computers are based on discoveries in physics and fundamental science that go back thirty, forty, even fifty years. What will come out of contemporary science—out of the research that we are doing today—we just do not know.

Question: In the light of your experiences, has the isolated inventor been generally the source of innovation, or is a new institutional development coming to the fore, the group of inventors?

Mr. Wiesner: One has the impression that in most fields the group has replaced the lone inventor. But it is my belief that more often than we realize, the original brilliant idea is still apt to be the product of one man's genius. He may, however, live in a group environment and have the advantage of all the inputs of scientific and technical competence and the intellectual interchanges that come from working with a large group of people.

At the same time I should emphasize that once a new idea exists, it takes many people to do the work necessary to transform it into a product. It is at this stage that innovation becomes a group and not an individual activity. Invention today depends not only upon a sophisticated body of information but also upon a sophisticated technology. As a result, building a vacuum tube or a transistor or almost anything radically new and important today is not a job for one man or even for a small number of men.

Question: In addition to the human factor, development also involves a number of other factors. For example, Farnsworth's discovery of the basic principles of television could not be put into effect until he had secured the use of several related processes that had previously been patented. These were eventually cross-licensed, but it was Farnsworth's opinion that the pace of innovation in this area was slowed down by this factor. Are delays of this sort a serious impediment to the practical application of a discovery? And if they are, can anything be done about it?

Mr. Wiesner: You can certainly find situations in which an individual inventor has a new idea that is useless without the application of prior inventions. Farnsworth's invention in the hands of one of the large companies, like Westinghouse, RCA, General Electric or any other large electronics company, would not have posed the same problem. The answer to this question could be given only if we knew how often a private inventor comes along with a basic invention. I am inclined to believe that the case you cite is rather special.

Another situation in which some inhibitions in applying a discovery may occur is when an invention would render a product obsolete that is already important to a firm. I do not know how often this has happened, but it is my impression that, for example, in the electronics industry today the cross-licensing of patents is so thorough and so accepted that new ideas are applied very quickly. It is possible to argue that because the Bell Telephone Laboratories had no interest in vacuum tube manufacturing and instead wanted the best devices they could get for the telephone system, they brought out the transistor promptly, whereas a tube manufacturer might have been slower. However, I doubt that this would have been the case.

Question: You have said that large organizations have an advantage because they can pool funds to collect people to work in research and development. Is there a possibility that the very same large accretions of capital that make it possible for the large firm to support research also tend to inhibit the exploitation of research? Does a large company have so many stakes in the market that it has to see a colossal amount of advantage in a new product before it will venture to disturb the old market?

Mr. Wiesner: Generally, only large companies are able to make the investment necessary to sustain a project over a long period of time without needing an immediate product or profit. Sometimes such an effort requires a large group. Sometimes it only requires a sustained effort by a relatively small group. It is certainly possible to find examples where the introduction of new developments and new products was delayed because of the fear that the new products might injure a company's markets.

Question: What efforts have been made to maintain intact successful teams of scientists and technicians so that they can be used in another area once a defense project is being terminated?

Mr. Wiesner: Not a great deal has been done, primarily because there has not been a lot of money available for it. In some cases firms have tried to do this for their own people. The Defense Department has helped to convert a number of groups, who used to work on airplanes, to research and development work on missiles. That kind of conversion is going on, but I think that as we stabilize the arms race, there will inevitably be some unpleasant breaking-up of groups of scientists and technologists. We do not see at this time, for example, the need for crash programs similar to those we had during the past decade.

A solution—and it is only a half-solution, because I cannot be too specific about it—is to work more intensively on the great social problems, such as transportation, desalinization, proper design of cities, and educational research, and thus to divert technical and scientific people into these fields.

Question: Could you expand on your point about industry in the Soviet Union? My impression is that most of their military products have been *created*, not copied, and I wonder whether this does not represent a significant achievement in terms of technological progress.

Mr. Wiesner: My comment was primarily about civilian industry in the Soviet Union. It is my opinion, speaking specifically about the military, that the Russians have done a remarkable job. However, even in this area, I think, they have depended primarily, though not entirely, upon Western ideas and innovations. For example, the jet engine they started with was, as I understand it, very similar to a British engine. Their metallurgical skills, however, were obviously first-rate, or they would not have been able to build those engines. For some reason, though, they have not been able to get their civilian industry working nearly as effectively.

THE CRITERIA FOR ALLOCATION OF SCIENTIFIC
RESEARCH EXPENDITURES

Question: What criteria should be used in deciding where our scientific research expenditures ought to go?

Mr. Wiesner: It is almost impossible to define rigid criteria. To estimate the relative value of research in high-energy physics or in genetics or in astrophysics is really impossible. Up to the present time we have been in the fortunate position of being able to support—sometimes well and sometimes poorly, and admittedly with some imbalances—all promising fields of basic research. And the term "support" may mean a few hundred thousand dollars for one field and $200 million for another. I do not know how anyone can establish absolute criteria that will lead to a conclusion that high-energy physics is or is not as important a field to support as genetics. One of the most important and most exciting fields today, as a matter of fact, is genetics, because it is asking some of the most basic questions. But the same is true of high-energy physics. It may be that we should use as a general criterion the question, how fundamental are the questions that are being asked?

Comment: I think that in speaking of allocating funds for purely scientific research you may have overlooked the fact that the process of innovation and development itself involves the need for more basic knowledge about how it takes place. We need to know much more about the processes through which the knowledge that is already available gets translated into technology that is useful to society.

Mr. Wiesner: I agree that this is a very useful and important area to explore, but I do not think that there is a need for criteria of choice in this area. The number of people asking for research support in this area is so small, and the amounts they need are so small, that one does not have to take funds away from physics in order to support research in the social sciences. In fact, one of the things that I tried to do in Washington was to stimulate social science research and to get more government support for it. But at no time in my experience in Washington was it necessary for me to choose between support for social science research and support of physical science.

Question: Suppose we came to the conclusion that the last thing society needed at this moment was further knowledge in the physical sciences, and that the real problem was to digest what we already know?

Mr. Wiesner: We do not propose to support high-energy

physics because we believe that society is immediately going to digest the new information and use it to do something productive. Our society bets on science because history tells us that scientific knowledge, in general, is likely to be useful sometime in the future.

Question: The federal government is now sponsoring educational research and development centers. People are trying to get support for educational research with the argument that because industry has advanced so rapidly through the allocation of large sums to research and development, education ought to do the same. Is this just a convenient analogy or is the relationship between research and development in education and the progress made in educational techniques similar to the relationship between research and development and the subsequent progress in other fields?

Mr. Wiesner: There are some lessons, which we have learned in industrial research that I believe carry over completely to educational research. One of them is that if we want to do a big job, we should not attempt it with meager resources. In the most successful educational experiments to date—for example, the curriculum development activities in the field of science— the most important thing that has been learned is the value of making revision and innovation the principal occupation of a substantial number of able people. Instead of a college professor who prepares his notes the night before class, and bases them largely on what he finds in other people's textbooks, an attempt should be made to hammer out what the subject is about, and then experiments should be made to find the best way of presenting it. I watched members of the Physical Sciences Study Committee develop a high school physics course. In the group were some of the country's most distinguished physicists. They argued for a long time about the definition of mass, and later about how to present it in a way that was coherent and comprehensible. It was interesting to see that even these unusually able people, once they had decided that they were going to do the job right, had to work very hard on a presentation planned for high school students. They spent months developing the right experiments for the Physical Sciences Study Course, discarding many—just as they would have done if the task had been to develop a good piece of experimental equipment—until they found some that they felt were just right. That physics course, extending only over two terms, cost $5 million. Incidentally, I would not call this research, although it may have involved some research; it really comes under the heading of development.

But there is also, it seems to me, much research to be done in educational theory and practice. There is a broad range of questions about learning which should be explored. How do children learn? What are the best types of examinations? How can examinations be used to reinforce learning? What are the best media for transferring information? So far, it has turned out that every time we have spent some money to learn how to teach better, we have found that we can improve teaching methods immensely. These experiences lead me to believe that improving the efficiency of education would probably pay most handsomely.

THE RATE OF INNOVATION AND THE PROBLEM
OF UNEMPLOYMENT

Question: Some people maintain that we have an unemployment problem because the accelerated rate of technological change is leading to a displacement of labor. Should the creation of unemployment be considered in determining whether new devices or advanced technology should be introduced in any particular industry, or should that just be ignored?

Mr. Wiesner: It definitely should not be ignored, but it is my conviction that there are much better solutions to the problem than delaying the introduction of more efficient machinery.

If we are not intelligent enough to take the measures that should be taken parallel with innovation in order to mitigate the human displacements, then we really do have a problem on our hands. But there are many things that are not being done that could occupy the unemployed. Our problem is one that Kenneth Galbraith brings out so well in his book, *The Affluent Society*. We have not yet learned how to make commitments to our common social problems. We do not pay our school teachers enough nor have enough of them; we do not build enough school buildings; we do not have enough hospitals; we do not have enough doctors; we do not preserve our landscape; and we do not spend the extra money or expend the effort to build beautiful rather than ugly buildings. I am sure that you could supplement this list indefinitely.

But to return to your point about a moratorium on innovation. To try to legislate a moratorium on technological change in a society such as ours, in which people cherish their freedom of action, would be extremely difficult. Moreover, I believe that the positive contributions of technology to our society and to human welfare are sufficiently great, and the problems of deliberately delaying the introduction of new technology so complex, that the sensible course of action is to try harder to learn to use technical developments constructively. Much more research in the field of social science is needed, before such problems can be tackled, and a better articulation of our goals and objectives is necessary, before we can make the choices that are implied here with any confidence that they are the correct ones.

Industry and the Paradox of Ubiquitous Individuation

by EDWIN H. LAND
President, Polaroid Corporation

My particular approach to nature and the world springs from an inherent love for and enjoyment in finding out how things fit together, how people interact with the devices they make, and how we make contact with the amorphous intricacies around us. By the time I was seventeen and a freshman at Harvard, this love for exploration had led me to my work in polarization, and a few years after that led to a laboratory in which other people worked with me. I had limited funds at that time, but I soon learned that I did not need money to do this work, because what induced people to join our laboratory was simply intellectual interest in the substantive content of the projects. Given a person with that kind of interest, I could take him, whatever his training or lack of training, and make life in the laboratory an enriching experience for him, and turn him into a productive person.

To this very day I still invite people from the production lines to join the laboratory, and I say to them, "I think this, I think that. I do not understand this, and that. Isn't this a puzzling result? How did that happen? I am going out for two hours; will you carry through? And if this happens, see if you can make it better." In a few weeks these people are working effectively, in a year they are valuable, and in two years I have a Pygmalion problem: they want to be independently creative! They create little worlds—perhaps all that any one of us possesses—and in these worlds they build, they generate, they come to understand certain things better than anyone alive.

I would remind you of the way Robert Oppenheimer has talked about the nobility and poetry of science; I can assure you that this sense of the nobility of science is not confined to the Oppenheimers or the Einsteins; it is available to all reasonably intelligent people. The dignity these people acquire from this sense of nobility, and the effect it has on the happiness of their families is something I rejoice to see. I have seen it happen to hundreds—I wish I could say that I have seen it happen to thousands. It is my dream that it *can* happen on a vast scale in times to come.

At the Polaroid laboratory we learned to work together and to enjoy life together while engaged in scientific activity. Our group has expanded again and again, but we have never lost that feeling of joy. I continue to find dozens of people in our laboratory who two years ago were on the production line and who are now excited about today's problems in science. It is my conviction that in the field of science there is no practical limit to the number of opportunities and places through which ordinary people (whom, incidentally, I regard as quite extraordinary) can enlarge their talent and expand their activity.

This is not generally what happens today. Where do we fail? I define failure here as a working life for many people in which they have the illusion of participating in the struggle with nature, but in fact are not; in which they are dealing with all the activities arising from the struggle and are concerned with its outcome, but are not deeply engaged, creatively, in the struggle itself and its delights of conquest. The locus of failure seems to be at the point where we reach beyond pure scientific research into engineering, and then again where we reach beyond engineering into production.

Our dream at Polaroid itself reaches very far. Thousands of people on the production lines are imbued with the idea that each person in industry should be as much a man, or as much a woman, when at work as when at home. They are imbued with the idea that the universal goal is the dignity of each person.

For centuries men could survive with the *illusion* that they were participating in creative activity. Illusions are not to be taken lightly, but now we are asking for more than illusions. In a restricted group of a few hundred people we *do* provide more than an illusion. One of the problems I have been working on, and I am speaking now as a scientist, is how to spread this wonderful sense of interplay with nature to a larger group.

The cause of our failure more fully to achieve this goal seems not to lie in an inability to pass on to the mass of people what we feel, because that seems to be easy; they are very ready to accept it. Rather, it lies in our inability to pass on to our managerial peers, the notion and the dream that they could do with thousands what we have done with hundreds.

For example, when we were still a small organization we would rent Symphony Hall in Boston for the whole day. We would have members of the Boston Symphony play some music, then have scientific lectures, and then we would have a review of the business activity of the company. We talked about our dreams and about what we were trying to make, and we also talked to the supervisors saying, "John, you are safe. Nobody is going to dismiss you if you take a chance. And don't be afraid of the people working for you. What we would like you to do would be achieve such a relationship that when you come in the people working with you say, 'Sit here, old boy, and guide us when we need help. We'll take care of this department's problems. We can do it.'"

It is clearly incongruous to place women, who run their homes with a great amount of incentive, energy, and efficiency, under the supervision of some young men just out of M.I.T. The young men tend to be so frightened that they keep the women strictly in their place. Yet there is not the slightest doubt that in the aggregate these women are potentially more constructive than the *individual* supervisor. The women could plan the work of their department and could have innumerable good operating ideas.

When it comes to the area of mass production, I urge that we need extensive cultural change, and it will have to be initiated by a group like this seminar. It must be a cultural change in which the pattern of life is no longer simply dominated by competition with one's peers. Until we have a culture in which we have re-examined the roles of aggressiveness, pride, and proper self-respect, no progress can be made toward a better society. Nothing we are proposing to do in a program directed against poverty or in a program to control the effects of automation can succeed. In this kind of cultural environment we are planning for one enormous hell. A world of peers whose sole activity in life is to take their magnificent intellectual endowment and use it *only* for their own benefit in the contest among themselves, while everyone else is left behind, is necessarily going to be a world of chaos.

And it is so *unnecessary*. Each of these people could have a trebly rich life while doing what he wants to do, while he is learning more about his field in all humility and with proper pride. While he is himself learning, he could find a way to bring along with him the hundreds of people in his own environment, who could both enjoy and share in his adventure.

In short, I think we have a choice in America today. We, who hold the responsibility for leadership, have a choice between a greedy, self-contained preoccupation with fighting among ourselves for superiority, or, on the other hand, a deliberate undertaking to develop a program for sharing the richness of our intellectual life with the mass of people around us. There are millions of people in this country with good, healthy minds, who are saying, "Please lead us."

I believe we can find people to generate so many new technological domains, that we can keep this country busy with all sorts of new products and services. If you have enough products and services you will find uses for them, and people will be kept busy in all the stages of research, engineering development, and production that will, in turn, result from them. Just what will be going on within the industrial framework in the future, I am not sure. Production may become so close to research that there will be only a small repetitive component left. We *can* keep people busy.

The real question is: How can we instill in people the wonderful delight in the intellectual examination of nature? I know that you can share that delight with any healthy person if only you can get him to participate. But how do we get enough trained people to train disciples? How do we expand this group?

All I ask now is that we examine this problem in the face of the fact that we do not know how to solve it. To examine it in the face of the fact that we scientists at present are drawing ourselves together and excluding the rest of the population from this aspect of our intellectual life, a terribly dangerous course.

There are those who say that we are undergoing such an enormous rate of change that our society cannot cope with it. They say that we must seek a change in the character of the entire social system if we are to have a society based not upon aggressive attitudes, but upon respect for the dignity of the individual. My answer to them is that I want to make sure that in the course of worrying about the minimization of work, we do not destroy a superb institution—our industrial framework —which was built for one purpose in the past, that of making products, but which in the future will have the second purpose of taking us out of our delightful, disorderly home life, where we cannot get anything done, and do not expect to, and of placing us for a number of hours a day in an environment in which there will be sequential accretion both of our own work and the work of other people. I maintain that industry knows how to accomplish all this, although it does not realize that this may be its most significant contribution.

DISCUSSION

THE TRANFERABILITY OF THE ACCOMPLISHMENT OF POLAROID

Question: What you have achieved at Polaroid seems to be a unique accomplishment on the American scene. What leads you to believe that it might be possible to transfer your accomplishment and the values that accompany it with any appreciable degree of success to the larger American industrial and educational scene?

Perhaps that question should be rephrased and cast in a different light. Taking into account the fact that the brilliant young scientists coming out of our universities have been caught in a mill of competition and have not been able to utilize all their faculties—intellectual and human—the question we should ask you is this: What made it possible for you, initially, to teach your scientists the kind of happiness and creativity that they have at Polaroid, and then transfer this feeling to the man on the production line?

Mr. Land: We set out many years ago to treat this as a problem to be solved like all scientific problems, and then to publish our solution, because, after all, there is no point in an isolated utopia. What was needed was the transferability you speak of, and you could not have transferability unless you could state what the problem was, and how you had solved it. So I would go to M.I.T. and Harvard, talking, much as I have tonight, and trying to persuade young scientists that this kind of life would be a good life. And many of these young people did come to Polaroid just because of these statements. But many of them were disillusioned, and that is why I am here to talk to this group tonight. I have to report that, in a sense, we have built only the suburbs of paradise.

The reasons why we have not had complete success are undoubtedly numerous, but the most important seem to be these: First of all, my technical peers were pretty happy with the *status quo,* which gave them interesting problems to work on. Secondly, much of their training was the training I have described—they were trained to compete amongst themselves.

If I were to start all over again perhaps I would make a more intensive effort to convert trained scientists. By failing to make those particular converts early enough I may have jeopardized the whole effort by leaving a gap in leadership. I have tried to fill the gap by giving commencement addresses in which I analyzed our problem as one of creating an elite that was not isolated from the mass of the people but which, instead, took responsibility for the general welfare of society. But there has been a failure in articulateness, a pedagogical failure, if you will, on my part; it may be that the sort of person who succeeds in his studies at a university, and who is competing in his profession, is someone I do not yet know how to reach.

Incidentally, I want to stress that the type of world I would like to see will not be brought about by Christian martyrdom, but, rather, by taking advantage of Christian opportunity. I have not sacrificed myself, nor have the people close to me who have made a good working life for the people on the production lines. Far from it! They have enriched their own lives, enriched the company's life, and by their efforts have made both the company and themselves that much more important in industry.

Question: When you say failure, what do you mean in respect to yourself? You have been successful in your company, as far as I can see. How far outside the limits of your own company have you tried to reach, and where is the failure?

Mr. Land: What I mean by failure is that my dream of twenty

years ago—a dream to create a corporation all of whose members acted in the way my colleagues around me in the laboratory were acting—has not yet been fulfilled.

Comment: I have had a parallel experience. In my early experience during the war, when I directed a large research laboratory, we took people off the production line to staff the
research activity, and the top people, the M.I.T. people, you
might say, joined in this very happy work. It was really a most
extraordinary situation in that compensation had no relation
to one's position in the hierarchy of a project. I think something
like this could persist, but only under a leadership that is eter-
nally novel in its approach. It is difficult to find people who can
keep this kind of situation going. We did not find enough of
them as the scale of our efforts increased, and as a result, after
a year or so, we reverted to the standard procedure. Perhaps
we could and should have trained more people to exercise this
kind of leadership; perhaps we did not succeed in releasing the
creativity, energy, and other qualities of enough people to carry
it on. We bogged down just at the time when we did not have
enough leaders for the number of workers we had. We started
with good leaders and things went fine until the projects
became bigger.

Question: Mr. Land has given us an example of an individual
who possessed a constructive social conscience, and who was
able to attract certain colleagues and make the lives of those
colleagues better than they might have been. This is clearly not
the first time something like this has been tried. There have
been prophets throughout history. The difficulty comes precisely when an attempt is made to extend the benefits beyond
the immediate area of the prophet. This, I should say, is in the
broadest sense of the term the problem of politics. If a man
has a political problem, as distinct from intellectual, religious,
or other problems, it seems to me that it is a question of finding
a mechanism that will somehow transfer the good that we find

in particular places to a broader field. This is my understanding of what politics is basically about. People interested in politics have been trying to struggle with the same question with which, if I understand you properly, you are struggling.

You seem to suggest that the industrial structure can somehow rise to this problem and serve as the agency by which it might be solved. I suppose that the historical analogue to this is the Church, which itself organized a society in order to spread the wisdom and love of the prophets more broadly.

Mr. Land: Could we change one word in your formulation? Could we substitute "framework" for "agency"? The implication of your remarks is that a little utopian island will usually remain small and inconsequential because history and politics will roll on in an unpredictable fashion. I have a rather different way of looking at this. If we examined the British Empire I think we would all agree that it has made a priceless contribution to civilization, whether or not people five hundred years from now wonder what the Empire ever did. By then it may be totally dissolved, but what it has done will still have been worth doing.

So I would like to think that our island, although little, will have extended consequences. There will always be such islands, but we should try to make each island into an island empire, with the sure knowledge that what is good about it will be assimilated for the good of the whole human race.

Comment: It may be that simply by avoiding martyrdom, as you say you have, you may have insured your failure. You cannot hope to be a spiritual leader and not be ready to face martyrdom. Also, I really do not see how your method could do very much to broaden society, because what you have described is really a kind of scientific monastery. If you examine a monastery, you find that it does not make any difference whether a monk prays or makes cheese: what counts is that he feels that his activity contributes something toward reaching a goal he believes in. This is exactly what happened at Polaroid.

Question: Mr. Land raised two questions about his experience. One was that although he considered that he was successful with a group of a few hundred people, he had somehow failed in conveying that same spirit to the rest of the organization. That raises the question of whether the organization is too large. But a more important question is brought up by his statement that one cannot really teach people about such things, and that instead an arrangement for some kind of apprenticeship has to be made. It is hard to understand why Mr. Land expects *us* to go out and do what he is suggesting, if he has not been able to succeed in demonstrating to his inner group of supervisors that it is important that they go out and lead the organization and become the nuclei for new organizations. Is there such an apprenticeship system within this inner group which leads some of them to go out and create satellite organizations?

Mr. Land: A few of my peers who did not feel themselves in any sort of conflict or competition with me were able to imitate my social experiment. But it is true that most of them were more preoccupied with being technically better than I. They did not understand what I was talking about in the domain of social experimentation. So only a few of the more secure ones were able to imitate me.

Did we set up an apprenticeship system for this purpose? I would have to say that in my innocence I did not know that it would be necessary. I took it for granted that my "family" would act just like me, and I discovered only in the course of time that they lived their own lives in their own way. The proper answer to your question is, why not set up such a system? Perhaps we know enough by now to set it up. If we can find people who are universal enough in their ability and sensitivity and empathy, perhaps they can be both distinguished in their field and act as leaders of the people as a whole.

Question: You have given us a case in which you have both

succeeded and failed. You have succeeded, at times, in communicating a sense of sharing in the joys of discovery and invention to a selected group, and yet you find it difficult to continue this process satisfactorily at all times. Perhaps your success in this function became a victim of your commercial success. As your success in communicating a sense of delight in the sharing of discovery led to effectiveness, productivity, and profits, responsibilities also came, and what is probably the critical factor—so did a certain amount of wealth and power—both of which are distractions. The feedback from these distractions to the initially creative spirit may have caused it to cease to be communicable on a large scale. I think that you communicated this sense of values and this joy not only by plan but also, and primarily, by conveying your own values through your own behavior so that, as a child learns not by what he is taught but by what he observes, the people around you sensed your goal. So I wonder whether this is not basically your problem: the commercial success you achieve by your functional success creates counterforces, which then sap the initial creative drive.

Mr. Land: These are certainly among the important reasons why we are not able to expand this spirit as rapidly as I could wish.

Question: The other day I was discussing Rousseau with my students. For Rousseau the answer to the problem you have posed lay in the establishment of a small community that would give people an opportunity for direct interpersonal relationships. A truism in sociology is that when you have a large structure it tends to become bureaucratic and impersonal. It seems to me that the basic problem is one of size. How can modern technology satisfy human wants in a mass society without sacrificing the advantages of life in a small community? Every utopian movement, every group holding the Arcadian dream, has come back to the idea of the small society. But, given the fact of modern, large-scale society, how far toward your goal can we progress? What kind of social and institutional structures do you have in mind, and how can you limit their size so as to avoid the transmogrification which would tend to frustrate your dream?

Moderator: I think the answer to that is implied in part in the title of Mr. Land's remarks, "The Paradox of Ubiquitous Individuation."

Comment: Mr. Land seems to address himself almost entirely, perhaps because so many of us are from the academic community, to the responsibility of the academic community for the kind of social experimentation he has in mind. I would like to suggest that perhaps one of the reasons for his success in the limited area of the scientists at the Polaroid Corporation lies in the fact that he had the advantage of working in a subculture in which the people who experience the joy of doing research also come to be looked up to by those with whom they come in contact.

The reason for your more general failure, however, is that in the much larger area of your effort, that is, among the nonscientists, the average person tends to be anti-intellectual. In the very large companies that I personally have known, when one does not talk to a fellow scientist, but listens, instead, to the nonscientists talking among themselves—people who occupy important managerial roles—the conversation is profoundly anti-intellectual. I think that unless you are able to do something to change the spiritual values of these people, you cannot expect to produce any lasting effect on them, and I would suggest that the only way you can accomplish this is to make the very top leadership of the large industrial organizations understand that this is what must be done.

I think that another reason why your experiments succeeded but did not spread throughout our society is that the problems with which you have worked have occupied a relatively short timespan. Therefore, you could prove your limited social experiment to be a financial success before extending it. If, however, the unit with which you are dealing is as large as, say, the desalinization process, you must expect much more failure.

Mr. Land: Your point that there is prestige in research and development activities at the Polaroid Corporation is well taken. I am not, however, convinced that there is an inherent anti-intellectualism in our society. What goes by that term seems to be no more than a protective mechanism, which people adopt so that they can feel significant in a society in which others have more prestige than they have. We have not found anti-intellectualism to be a problem at the Polaroid Corporation, except in the very initial stage of penetration. It only takes a day to change someone from an anti-intellectual to an intellectual by persuading him that he might be one!

SOCIAL EXPERIMENTS AND SCIENTIFIC EXPERIMENTS

Question: You said that you approached the social experiment that you carried out at the Polaroid Corporation as if it were a problem in physical science. Are you suggesting that the problem of politics in the large sense of the term is susceptible to the methods of the physical sciences?

Mr. Land: When I say that my approach to social problems is a scientific one, I mean that we can learn certain things from the methods of science. One of the reasons for the rapid growth of science is that somehow or other people learned to free scientific experimentation from involvement with problems of theology and from the sense of guilt. To me the most impressive thing about science is that there, and there alone—I know of no other such place in life—one can fail and fail and fail without a sense of guilt. We must learn how to transfer this attitude to experimentation in the social sciences.

Therefore, I feel that in our social experimentation we can learn from science how science set itself free. I know the enormous vulnerability of each of us. When I come before you and say some of the things I do, what goes on in your minds a little below the conscious level? I must use all the credit I have in order to dare to talk about the things I want to talk about, and still maintain your respect. If I were to talk to you about a problem in physics, there would be no such dilemma. In any discussion of a problem in the social sciences, however, the very first thing that is apt to happen is that people look around and wonder, "Is he a fool? Is he an eccentric? He may have done it somewhere else, but he can't do it here. He cannot transfer success in one domain to another domain, where nobody can succeed." Anything we try to do, even in discussion, is immediately suspect however modest it may be, whereas in scientific areas we can do the rashest things and simply increase our budget for them.

What I am proposing is that we need a calculated and continuous program of experimentation in the social sciences, which will yield an enormous payoff if it succeeds, but which will not endanger anything in particular if it fails. This is how science approaches problems, and that is what I mean by using a scientific approach to solving social problems.

Question: There is a good deal of experimentation going on in many industries to capture something of the spirit you have been describing. An example that occurs to me is a very large company which attempted to reorganize the workers on the assembly line into small groups of people who would put the whole product together themselves. The experiment was carried out for a considerable period of time with a great deal of resistance from the workers. They did not want it from the start, and some even contrived to get a medical excuse to be released from participation. At the end of the experiment the girls were asked, "Would you be willing to continue with the experiment?" A few said, "Yes," a few said "No," but the great majority said, "We would if everybody else worked this way, but if we are the only ones who have to work this way, never! We are not even sure that if the entire company were organized this way we would want to keep it up, unless our husbands also worked in a company like this, because right now they think we are crazy!"

What is it in our environment and our social structure that creates attitudes that seem to make it much more difficult than you have indicated to get people to accept the utopian changes you have suggested?

Mr. Land: This experiment, which was discontinued, was called Job Enlargement. What was the difference between it and what I am talking about? The mere diversification of a series of mechanical tasks, although it seemed like a revolution at the time, is not the intellectual and emotional equivalent of a person's participation in the pursuit of the unknown. What is exciting for the people I am talking about is using this afternoon's learning and accomplishment for tomorrow's investigations.

That is quite a different thing from expanding the domain of one's training in the performance of the repetitive work of some task. I am not dismissing the experiment; it is just the kind of thing we want, an open-minded comfortable experimentation with possible solutions. It conforms to the slogan, "Evolution is achieved by revolution within restricted areas, where the ratio of probable gain to probable loss is high." I am very much for trying something quite radical, but in a situation where, if it fails, you have not lost very much, and, if it succeeds, you have gained an enormous amount. You have not jeopardized the organization, but if your proposal works you have learned something that you can then generalize from.

If you compare the Job Enlargement experiment with what we have tried to do at Polaroid, you will find, first of all, that if you put the average worker into a research and development laboratory he may never want to go back to where he was. In the Job Enlargement experiment, though equally interesting as an experiment, you expand a person's job; he has just been given a little more responsibility, but apparently no spiritual reward for the extra responsibility. The problem, it seems to me, is to set up a large number of such experiments—not one too risky in itself—where each successful outcome is treasured and saved as a lesson.

Question: I agree very much with what Mr. Land has said. There have been many sporadic attempts to create a job environment that was enriching and creative, and many of them have failed. I think that Mr. Land has put his finger on the root of the trouble. The experiments involved merely an enlarging of a job in the physical sense, and the workers, for the most part, perceived them to be merely as one more device to increase productivity or to improve quality. The job changes were not initiated (I am speaking from the workers' point of view) as something creative and enriching. They were imposed from above, and this raises a very profound question: What is the end of work? To the worker the end of work in our industrial society is profit for the organization. It is, therefore, very difficult to convince people that the kind of work you are talking about in programs such as Job Enlargement is valuable in its own right, not necessarily because it leads to greater productivity, better incentives, better bonuses, or what have you. Unless changes of this type can be related to something that will show up as something other than the usual ends of work itself, I am afraid that they are doomed to failure.

Mr. Land: At Polaroid we have a slogan, "The second great product of industry is the fully rewarding working life for every man."

MOTIVATION AND TECHNIQUES OF REWARD AT
THE POLAROID CORPORATION

Question: Given your aversion to competition, when, how, and why do you reward the technical people in your own company?

Mr. Land: All of us who have professional training are accustomed to think of excellence as much in terms of competition with our peers as in terms of achievement. It is this to which I am objecting. However, life is so homogeneous that dissecting the reward system into two parts is unnatural. The rewards are knit together. You get certain rewards from your peers. All of us respond to many kinds of rewards: appreciation, understanding, insight, money, position. As it stands today the only reward system for excellence in our group is a reward system that involves the rest of the group. It is only what we think of each other that counts for us. We have no world except our own world, so I cannot tell you how you might reward people outside our world.

Question: Do the people in your laboratories have such an ideal environment that they would come to work even if they were not paid, particularly if they had an independent income? Is it money that motivates them, or is it something else?

Mr. Land: We try to pay our people enough so that money will not be a chief motivation, but merely an incidental one. We pay enough so that people can live a happy and healthy life, but we hope that their primary motivation will be their interest in their work. Of course, there are times when money helps in scientific work. These are the times when a man is lying fallow, as we call it. That can last for quite a few months, during which there isn't any reward except memories of past successes. If you fear that something has happened to your creativity, and you think that you are not going to have any future successes, money is often nice to have.

Comment: There is a great gulf between Mr. Land's dream of creative, ennobling work and the reality of work. We must do much to change the reality of work before we reach the point where it will be possible for work to become ennobling and enriching. For example, it seems to me that we would have to do away with punching-in at eight o'clock in the morning, before we could begin to talk about a broader and higher goal.

Question: Is it possible to achieve the dominance of a feeling of generosity toward others, or is the biological mechanism with which we have to work and which we are trying to train only capable of being trained to show such generosity and love sporadically?

Mr. Land: I am saying that people do not have to be generous very hard or very long; they just have to be shown how to be generous naturally. They have to be trained to exercise this faculty. At the same time, one must maintain a channel of aggressiveness for the healthy functioning of a social system in which the channels of vigorous aggressiveness are not destructive ones. The most useful kind of aggressiveness is man's struggle with the forces of Nature. What a brilliant idea, for example, it is to capitalize Nature. By putting a capital "N" on the word nature, we have made Nature a victim, yet we have not hurt anyone.

The Management of Change: A Personal View

by ROBERT L. HERSHEY

Vice-President, E. I. du Pont de Nemours and Company

WHEN I AGREED to speak before the seminar, it was arranged that I would talk on "The Management of Change." I have decided to make the title a little less formidable by focusing only on personal experiences, because I am not sure I can competently talk about the management of change in the broadest sense. I can share with you, I hope, some experiences that have been enlightening to me in managing some changes during the thirty years I have been connected with the Du Pont Company. The conclusions I have drawn from my experiences about the management of change may help us make some generalizations in this area.

Before I talk about my own experience, I must say something about the atmosphere, the milieu, of an industrial organization like Du Pont. There are certain preconditions which determine what we do and what decisions we make. I am going to mention a few of these. They may be very obvious to you, but I shall mention them so that we will all have them in mind.

First of all, the Du Pont Company has to be run in such a way as to make a profit. We must take this into account in everything we do. This does not mean that everything we do has to be profitable at every moment. But the enterprise as a whole, over a period of time, must show a profit. I do not mean to belabor this point, but I think that you should understand that it ranks first in my mind as a necessary condition for the operation of the Du Pont Company.

Another thing, the implications of which are not, perhaps, so obvious, is that we assume that the Du Pont Company is going to be in business as far into the future as anybody can see. We assume that we are operating an organization that has permanence. This puts a very great restriction on some of the decisions we can make. We cannot make a decision just by looking at what results it will bring next week; we must look ahead to see what the probable results will be next year, or in the next ten years. We hope to be able to foresee some of the things that might happen. Very often, though, we cannot see too clearly.

In my view, American business history is littered with the mistakes made by shortsighted managements who did not look down the road to determine the long-range consequences, before making decisions on current problems. The consequences of these mistakes have been most unhappy for the individual companies and for American society as a whole. The point I want to make is that, with no claim at all that we do it perfectly or even very well, we do try, in the Du Pont Company, to look far enough ahead to see that we are not making an easy decision today that will create serious problems in the future. If one is simply running an operation in the hope of selling out for a nice capital gain, one can make very different decisions from those we are permitted to make, or should make, for a company we assume is a permanent organization.

There is also, of course, the constraint of competition. I would be less than frank if I did not say that we try, by all legitimate means, to lessen the impact of competition. The usual tool for this is to try to achieve an exclusive patent position. If one cannot get a patent position, one tries to get a dominant market position so that most people are buying the company's product. But believe me, I do not know of anything we do that does not meet some competition. We have to be out there in the market giving the customer what *he* wants, at the price *he* will pay. And we have to achieve a balance between cost and price that will make us some money.

In short, the experiences with change, which I propose to share with you, always took place under conditions in which the necessity for securing an ultimate profit was of first importance; to secure that profit, consideration of course had to be given to the competitive situation; in addition, the decisions taken were qualified by our desire that they be of long-range as well as short-range value.

Let me now sketch out for you some of the situations at the Du Pont Company in which I have been involved, in one way or another, with change. Some of them were situations in which the decision to make the change lay wholly within the power of the management. Some of them were situations in which external influences forced the change, and the question was then one of adapting to what the external influences were forcing upon us. Some of them have to do mainly with what I myself believe to be the most crucial factor in change: people, and not things.

I came to the Du Pont Company in 1936, after having been in the chemical engineering department at M.I.T. for about twelve years. I went to the Du Pont experimental station outside Wilmington, to be what was called a leader of a semi-works group. In this job I had the responsibility of planning and executing, with the assistance of other engineers, the things that had to be done in order to collect enough information and data about a discovery an organic chemist had made in the laboratory to be able to tell the design engineers, "Here is the way to use this organic chemist's discovery in an operating plant." I did not have very high-level decisions to make at that time. The management had already decided what particular project they wanted to undertake. It was my job, as a technical expert, to take a scientific discovery and carry out the first steps toward utilizing it in an operating plant.

My very first assignment involved a chemical reaction that took place at very high pressures. In those days, the only other operation that was run at such high pressures was the synthesis of ammonia. The chemicals we had to handle formed an extremely corrosive mixture. At the temperatures required, steel was dissolved almost instantly by this mixture, so the chemical reactors had to be of steel for strength, lined with silver for corrosion resistance.

I must confess that I myself do not know where the line should be drawn between science and technology, but I suppose you could argue that the technology, in this instance, consisted of knowing how to construct the reaction vessel for this corrosive mixture and how to make the flanges and the gaskets at the ends, so that they could hold the mixture. This is what we had to do. Furthermore, we had to explore this reaction over a range of conditions in order to find out the optimum conditions for production, not in a technical sense, but in an economic sense. The problem was how to build a plant so that the investment would be justified by the earnings. In this case the market already had many similar products in it. This necessitated a study of technical alternatives, and before a specific design could be adopted the economic consequences of the technical alternatives had to be compared. Ultimately, we came up with some design specifications. I thought, in those days, that all we had to do was to produce these specifications and hand them over to the design engineers. They would then apply their ordinary design techniques and hand over the finished design to the construction people, who would build the plant, which would then be run by the operators.

Well, the first two jobs I worked on brought poor results, very poor results, once we got them into the plant. We had more trouble than you can imagine to make the process work.

And the only claim I can make for myself in the whole matter is that I was smart enough to learn from my sad experience. So each successive time, we tried to do something a little differently. Over the period of seven or eight years during which I was intimately associated with this kind of work, we must have done a dozen of these development projects. And when we looked back at the early experiences, it was quite obvious that the root of our trouble was that we had not handled *people* correctly. We had waited too long, in our development work, to get the point of view of the design engineer; we had waited too long to get the advice of the construction man; we had waited too long to get the suggestions of the operations man.

And so, in the later projects, we started bringing in the people who were going to have responsibilities further down the line. Instead of making a sharp cut-off, we would bring them in early to make them familiar with the process. They had valuable suggestions, too, about their special problems, and I think we became a good team by the time we were working on our third or fourth project.

These were very educational experiences for me, but the particular scientific discoveries we made were not, in the long run, revolutionary. As a passing commentary on technical change, I might say that, though they served their purpose for a time, the plants we built for the first two processes have long since been abandoned.

Perhaps a better example of the flow of technical change is found in the chemical intermediates for nylon, a project on which I also worked in the 1930s. Two main chemicals go into the manufacture of "66" nylon. We had to develop the processes for making these chemicals from scratch. Ten years later we had completely replaced the original process for making one of these chemicals with a wholly new process. This had not been thought possible at the time the first plant was built; we had also greatly modified the process for making the other chemical.

The real point I am trying to make about this experience

is that we found we had to get different points of view on these problems; we had to get coordination and communication among people; we had to learn—if I may say so—how to *manage*. I am reminded of one more aspect of this problem. Some years later, when I was further away from this kind of development work, the same department had an experience that made me suspect that we had lost the technique of management I thought we had learned: I took time to investigate, and it was clear that the people involved did not know the technique. The reason for this was that they were not the same people who had learned it the hard way, years before. This was another generation, and they had to relearn what we had learned. There is less danger of this happening today, because formalized methods of planning are being developed, so that people take into account the things they once used to overlook. But it should not come as a surprise that acquired skills can be forgotten; they have to be relearned by each generation.

I would now like to touch on two later experiences, which are quite different, and which resulted from problems encountered when I was in the general management of the polychemicals department. In those days we had a very wide range of products in the department, but we did not have a proper sales organization to get those products to the market effectively. Everyone in a responsible position in the department realized that we had serious problems. But each of us (we were all individualists), had his own ideas about what ought to be done. And unfortunately, the ideas did not coincide.

Finally, we had a consulting firm come in. They did not have any new ideas, either, but they were a very good catalyst for bringing people into agreement. They made a very thorough study of our operations, and, being outsiders, they could say things quite frankly and pointedly to people to whom those things needed to be said. And they had the courage to say them. That was the way agreement was finally reached.

But what I am proudest of, in that operation, is that there was not a single man in the organization who woke up one morning and discovered that he had a new job by reading it in a newspaper or on a mimeographed sheet. We were very careful to tell everybody, starting from the top, what our objectives were to be in the new organization, and what part we wanted each person to play in it. They may not all have been satisfied—I am sure some were not—but there was not a single surprised man in the group. The purpose of all this preplanning, of course, was to ensure that everything would go smoothly the day after the changes had been made; that the new approach would receive understanding, acceptance, and enthusiastic support from the people involved. I am happy to say that all happened as planned.

The second instance I am going to touch on involves a plant in Arlington, New Jersey. Around the time of World War I, it produced toilet sets—mirrors, combs, etc. By the end of World War II, this plant had been reduced to a one– or two– product operation; all the other products had become obsolete or had been moved elsewhere.

It had one operation, however, in which the product was not obsolete, but the process was. And it was impossible, physically, to put the new process—which, incidentally, we had not completely developed at that point—into that plant. At the management level we had two choices. We could go out of the business, or we could make a large investment in developing a modern process, putting it into a new site, and then facing up to the problems of what to do with the employees at the Arlington plant. We had the courage to take the positive and not the negative action, but, from the time we made that decision, and for the next three years, we lost money on the product. There is no mystery about this product; it is the laminating layer in automobile windshields, which holds the two sheets of glass together.

The most difficult problem was what to do about the Arlington plant. If we took this product out, we could no longer afford to produce the remaining smaller ones at that plant. The plant had, at one time, employed over 3,000 people; at the time in question it was down to about 500. The polychemicals department, however, had never faced the problem of completely shutting down a plant. Indeed, in the whole company, there had in recent years been only one or two partial closures. We had to consider what we were going to do with these people. We certainly were not just going to throw them out into the Jersey marshes. Many of them had served the company a very long time. At that time we did not yet have formalized plans for dealing with this kind of situation.

We went to work with the employee relations people to develop a policy that would not only take care of the Arlington plant, but could also be adopted as a policy guideline in all similar situations for the company as a whole. I think we came up with quite a good solution. It included some flexibility in the company's long-standing pension plan, and provided severance pay schedules that were generous and proper. And then, the first moment we could, we began operating an employment office to help those men get other jobs in the area. Thus, we phased out this plant, not solely on the basis of what it would do for us, in a money sense, but also on the basis of what we could do to preserve the company's reputation for dealing properly with its employees. The whole operation, I think, went very well, but we were not very lucky in transferring those people who were paid by the hour to jobs that were any great distance away, because they were not mobile people. That is, they were not able to go into a new community and establish themselves satisfactorily there.

Let me just mention a few conclusions that I have reached about this business of change. First of all, a study of the objective to be achieved must be made, and the situation around that objective must be looked at in depth. Secondly, the obstacles that stand in the way of achieving that objective must be considered. Sometimes these obstacles are things in the physical world. Much more often, however, they are things in the psychological world. Thirdly, the question must be asked, "What is the priority in which I will try to overcome these obstacles?" It is no use trying to overcome the last one if the one that is really the first has not yet been eliminated. And, finally, a great deal of persistance and an infinite amount of patience is necessary.

I am a drop-of-water-at-a-time man. I have seen too many very keen, smart young men rush in and say, in effect, "This is so obviously reasonable and correct that anybody who doesn't buy it is a stupid fool." Unfortunately, the young man is often the stupid fool. He rushes in and at the very outset creates an adverse emotional reaction; then the progress that he makes, after a long time, only brings him back to where he started.

My own experience suggests that changes can often be effected soundly and quickly, if, when rebuffed at the outset, one refuses to go to the mat with the opposition for a decision. Instead, one patiently waits for the next opportunity, and then uses it, and the next, and the next, to persistently make one's point. I call this the "drop-of-water technique." No single drop does very much, but the cumulative effect is decisive.

DISCUSSION

CRITERIA FOR ALLOCATION OF RESEARCH AND
DEVELOPMENT FUNDS

Question: Could you give us an idea of how Du Pont decides where to put its research funds?

Mr. Hershey: Over the years we have had many different ways of classifying our research activities. We have had classifications called "fundamental research," "new process research," and "new product research." About five or six years ago we took the first step toward classifying research expenditures not by the character of the work, but by its business purpose. The main reason we do research at all—let us be frank about it—is to further business.

About two years ago I became the executive committee advisor on research—we do not really have anybody "in charge of" research. Each of our industrial departments, there are eleven of them, has its own research group to serve its needs. We also have a central research department that concerns itself with long-range projects. In addition, we have a fairly large research division in our engineering department. Finally, one of the members of this seminar, Dr. Edwin Gee, supervises our newest corporate research venture, a relatively recent addition to the development department. In short, we are a very diverse company with a very diverse research set up.

When we began studying our various research departments and activities, we found that they were not pinpointing what Du Pont's industrial (money-making) departments needed to have pinpointed. We spent about nine months preparing a proposal for a new policy statement, and we finally decided that our new policy statement should begin with a reaffirmation of the fact that the future of the company absolutely and critically depended upon a vigorous research and development program.

The first business purpose of a research and development program is to keep our present products fully competitive and, if possible, better than competitive. Whatever results research and development work produces are thrown into one hopper and labeled "improvement of present businesses."

Then there is another kind of research that we have to do, which is oriented toward the future—we have to find leads to something new. This is, if you wish, what people call "basic research." But I would like to point out that very often in the course of improving the position of our present products some

basic scientific work of very high quality is done. From that point of view there is no strict differentiation between the basic scientific work that is done for its own sake, and the work that is done for business purposes.

To encourage the finding of new leads, we set up a category of work which, for want of a better term, we called "exploratory research." This can be almost anything: the most searching, the most extraordinary, the most uncertain (in terms of practical outcome), and the most difficult type of research.

And then, when something rears its head in this area of exploratory research that looks as if it is going to move on and become a new business venture—it is generally at this point that the dollars begin to flow out fast—we reclassify it as a "new business venture." From that point on we accumulate all the charges against it—technical, commercial, market surveys, in short the entire cost of the thing—and report regularly on it to those people who will have to appropriate the money to build the plant.

We have only just begun to accumulate figures in this fashion, so our experience is not very extensive. But I would hazard a guess that exploratory research, out of a total research budget of approximately $100 million, receives about $10 or $15 million.

The $100 million, incidentally, includes research and development and all the collateral expenses that go with it. If bringing out a new product is to be a success, one must, in most cases, spend very substantial sums to answer such questions as, "Can I sell it?" "At what price?" "How large a quantity?" These are not technical questions at all, but they do have technical aspects because the answer may be, "No, I cannot sell it in its present form, but if I make just a small technical change in it, then, indeed, I can sell it." So, returning to the laboratory, one figures out how to make this change, reassess what it is going to cost and off one goes again to the marketing department.

In the last analysis, research spending is just a matter of

judgment. The most difficult job in the business world, in my opinion, is managing the real research part of a research program. The results of a sales program or of a manufacturing program can be measured; within a year or so one can tell whether the program was successful or not. But to assume that one can tell whether a man will discover or invent something between now and the end of the next six months would be crazy. If it were predictable, it would not be research. And since research is unpredictable, managing it is a problem.

I do not think that dollars are at all a meaningful way to measure research effort. During the years I have been an advisor on research at Du Pont, I have met a number of economists who wanted to make a study of the economics of research and development. They seemed to have the naïve idea that there is some quantitative formula for measuring research. I can assure you that there is none. I could put together a research organization that would be guaranteed to spend twenty million dollars a year and not produce a single thing. And parenthetically, I think we already have some that do just that.

Question: But you do make a decision to spend a $100 million each year at Du Pont on research and development, not $200 million or $50 million. Therefore you do have some criteria, however undefined they may be.

Mr. Hershey: But I must emphasize that the decision to spend $100 million is not a decision made at the top. It is the cumulative result of decisions made by people who have to make the day-to-day business decisions. It is the responsibility of the general manager of a department to show a profit. He looks at his products and at the competitive situation, and he and his technical or research director sit down together and say, "Here something can be done to cut costs. If we knew more about this chemical reaction, we could carry it out in a cheaper vessel, or we could make the reaction go faster, or we could do something else." They explore in detail what the problems of that particular business are. Then they say, "We want to get this particular problem solved. So, for the next year, as a business decision, we must spend this amount."

Question: If, at the end of the first year of a project, you have lost money or the research does not work out, what do you do then?

Mr. Hershey: Then we reconsider the situation and start all over again. We make decisions like these one at a time. We would be insane if in the top management of Du Pont we were to say, "We are going to spend a $100 million on research; now you boys go ahead and chew it up."

TECHNIQUES OF MANAGING CHANGE: THE PROBLEM
OF LEADERSHIP

Question: You stated that the management of change deals with the management of people and with problems of communication and coordination. How is this management of people carried on in the various areas concerned at the Du Pont Company?

Mr. Hershey: The man at the head of the organization sets an example that is followed right down through the organization. If he conducts himself in such a way that he sets the proper example of communication, by and large the organization will have a free flow of communication, even though there may be some blockages here and there.

Question: Dr. Land of the Polaroid Corporation has told us of the necessity of making the average production line employee enthusiastic about innovation and invention. Does Du Pont consider this essential, and if so, how does the Du Pont Company instill this enthusiasm in its employees?

Mr. Hershey: Whether employees are enthusiastic and creative in their jobs in my opinion, depends on the leadership they are getting.

Question: Could you give us a particular example of a case where you stop thinking of improving a product and start thinking of finding a substitute for it or of abandoning it?

Mr. Hershey: As a general rule you can say that as long as the company keeps recouping its costs of production and makes some profit, a particular item will go on being produced. If a point is reached at which the market ceases to grow, or is beginning to die, it becomes more and more a question of how much additional money the company should spend on facilities for the production of this item.

Let me give you a specific example. Growth in the cellophane market has become less rapid in the face of strong competition from a great variety of new polymeric films that have come in as a result of scientific discoveries and engineering work. There is very little realistic expectation that cellophane will ever undergo a rapid increase in sales.

This is the kind of situation in which one wonders, "How much can I afford to put into the further development of this product?" This is simply a matter of judgment, as is the question of research spending. I might call your attention to an article in the January, 1965 issue of *Fortune* entitled, "This Monster, R and D," which pointed out that many industrial firms are becoming more and more selective in their research spending. There is no truth to the supposition that merely spending a lot on research is going to bring any money into the till. It may turn out that the discoveries are unsaleable.

Question: In looking at managing during this past decade, a period in which, many assert, there is an increasing rate of change, do you see any differences in the type of young people that you bring into your organization? Also, has this had any impact on Du Pont's internal education and training programs? What is your evaluation of such training programs?

Mr. Hershey: I should say that we have done more training in the last fifteen years, and I think we generally have to do a great deal of it. Mr. Gee, the Director of our development department, can tell you more about this.

Mr. Gee: For production employees there is a great deal more education today, as far as shifting skills is concerned, than there was in the past. On the other hand, we do not give a great deal of training within the company to men with technical training.

Question: Assertions have been made that as many as 10,000 engineers may be becoming obsolete each year because they have been trained in a specialty that has lost its usefulness. I suspect that in many areas this occurs to people whose background should make them easy to train and to use.

Mr. Gee: I think that the basic problem here is that in many cases these people lack an economically usable training. They may have excellent formal training, but they do not have practical training. The utility of a person who has been displaced from a government project to a commercial project is apt to be marginal because he is not accustomed to think in terms of the markets, of costs, and of selling prices.

Mr. Hershey: This is also a question of the rate of change. In our own experience at Du Pont we have often been able to deal with the problem of obsolete skills by letting attrition work. We do not rush in too quickly in cases where, if we wait a little longer, the natural rate of attrition will solve the problem. This method has not produced any major crises in the company.

SMALL SCALE VERSUS LARGE SCALE CHANGE:
TOTAL MANPOWER NEEDS

Question: Up to this point in the discussion we have been concerned largely with the management of small-scale changes. We are, however, in the midst of an era of very major change, a change in which manpower needs will be drastically reduced within a few years time. Have you done any planning to meet a situation in which total manpower needs will decrease?

Mr. Hershey: I believe that the drastic reduction in employment per unit of output, that we have had since the war, has been slowed in recent years.

Question: The growth in productivity over the last thirty years indicates that the derivatives of growth do not always take the form of a reduction in manpower. Of course, these derivatives may change in the next decade. There is no economic indicator that says that they will remain constant.

Question: Do you think we can put in more and more automation without reducing manpower needs? If productivity increases every year by a certain percent, conditions in the foreseeable future will force us either to cut down on working time, or to push more and more products down the people's throats.

Comment: Another way of putting that point is to note that this awesome increase in productivity is increasing the standard of living. Take the case of a person who has four Tchaikovsky symphonies and then gets the opportunity to buy five Beethoven symphonies because his income has gone up. I do not know whether this is forcing consumption or whether it is raising man's cultural level. I would suspect that it is the latter.

THE ROLE OF THE LARGE-SCALE ORGANIZATION IN SOCIETY

Question: Mr. Hershey has described the process of making decisions about particular products in many illuminating ways. He has emphasized the profit-making purpose of such decisions and the preoccupation of the very large firm with the future effects of current decisions. It seems to me that much of what he said could also be said of a small organization. I wonder whether an organization as large as Du Pont has a conception of the future of society and of its role in that future which differs from that of the small entrepreneur. Particularly, I wonder what Du Pont feels its role will be in a society in which there is an increasing fusion of government and private enterprise. What weight does its preoccupation with its future role in society carry in Du Pont's planning, in the kind of products it develops, and the approach it takes toward their development?

Mr. Hershey: I have some trouble responding to your question. You ask me what we see in the future, but I have no better crystal ball than you. Can you tell me what bills are going to be passed in Congress in the next five years? I cannot tell you. All I can say is that we look ahead, but we do not know enough to make rigid predictions or to plan in an inflexible manner.

Question: On the other hand, you have suggested in your talk that there are a number of things that the company does that involves planning for the distant future.

Specifically, to what extent does the Company anticipate situations in which it will have to close down plants during the next ten years?

Mr. Hershey: We now operate in some areas in which we are quite certain we will have to shut down plants. The department head responsible for a particular business area has a preliminary schedule, subject to constant revision, to enable him to plan what to do with the people involved and to determine whether it is possible to put another industry on that plant site. This we have succeeded in doing in some instances. For example, we have a plant in Tennessee that originally made viscose rayon and cellophane. It doesn't make either today, although it still has very many employees. This was made possible by putting in other operations that were suitable to the plant, and, in the meantime, phasing out the old ones.

There are always dying products, and obsolete processes that have to be eliminated. We try to foresee this, perhaps as much as five years in advance, and begin making plans as early as that, so as not to upset the surrounding community when we shut down.

I do not want you to get the impression that the Du Pont Company has a magic solution to the problem of the consequences of a changing technology, changing markets, and what have you. We have tried to foresee such changes, and up to this moment, I think, we have been fairly successful.

Question: A number of your statements about the decisions made by Du Pont imply that they have a considerable impact on society as a whole, for example, the impact on a city of a decision to close a plant.

Also, you have said that mistakes are sometimes made by Du Pont's management. Since Du Pont is such a large organization and any mistakes are, therefore, going to have a tremendous impact upon local communities and on various aspects of our national economy, perhaps some sort of political perspective is necessary for planning.

Mr. Hershey: But you make an assumption about central economic planning that I cannot accept. What makes you think that politicians will not make their share of mistakes, or perhaps even more than their share?

Comment: But at least they have a different perspective to bring to these problems.

Mr. Gee: The foregoing comment overlooks the single best answer for the problems to which it refers, namely, that Du Pont and other large corporations be healthy, profitable, viable enterprises. If Du Pont, for example, had not automated, if it had not had an aggressive research program, the company's export sales today would not be the equivalent of the total sales twenty years ago. Think what a contribution that achievement makes to the solution of our balance-of-payments problem!

Moreover, the reason why Du Pont's employment is currently increasing is that Du Pont is a healthy, profitable enterprise. It is just because it is a profitable enterprise that it can take care of those people who are displaced when plants shut down.

Mr. Hershey: What you are asking is that private industry assume more social responsibility in its own enlightened self-interest, though legally its only responsibility is to its stockholders, and to its employees, and to a few other very limited segments of society.

Question: This reminds me of an essay published in a recent issue of the *Columbia University Forum*[1] in which the question

---

[1] James Kuhn, "What's Wrong with the Old Business Ethic?" *Columbia University Forum*, VII (Summer, 1964) 22-26.

was raised as to whether the U.S. Steel Corporation, as one of the primary determinants of employment in Birmingham, Alabama, does not also have a certain social responsibility for decisions taken in the name of its shareholders, ostensibly with the sole intention of maximizing profit. Every large corporation must face the wider social consequences of its actions. My question is, simply, to what extent do management decisions take this into account?

Mr. Hershey: I have tried to emphasize that this was one of the things I had in mind when I said that we looked upon the Du Pont Company as a permanent organization. I tried to make it clear, by giving some examples of change, that we feel very acutely a responsibility to the communities in which our plants are located, and we have initiated very elaborate, careful, and relatively successful programs to meet that responsibility.

LESSONS FOR THE PRESIDENT'S COMMISSION ON AUTOMATION

Question: If you were to take the experience of Du Pont between 1920 and 1940 and compare it with the period between 1940 and 1960, would you say that there has been any substantive difference in the general character of the second twenty-year period compared with the earlier period, which might offer some lesson to the country and which, therefore, ought to be on the agenda of the President's Commission on Automation?

Mr. Hershey: In the proceedings of one of your previous meetings[2] one of your speakers attempted to relate technological progress to the degree of concentration in industry, using as examples, among others, the chemical and the steel industries. I do not think that concentration is the proper independent variable in this case. Rather, it is the quality of management. The steel industry, like many other industries in

[2] See the essay by Alvin M. Weinberg in, *The Impact of Science on Technology,* ed. by Aaron W. Warner, Dean Morse, Alfred S. Eichner (New York, Columbia University Press, 1965), pp. 63-74.

the earlier industrial development of America, was an industry that initially had no scientific basis at all. The production of steel was an art that had been modified by ingenious inventors, and it was not based primarily on scientific discoveries. This situation has tended to persist. In my opinion, the reason why the American steel industry, by and large, has not been more progressive is that until recent years its management has largely been made up of lawyers; scientifically trained and technically oriented men have not had leading positions among its top executives.

To turn more specifically to your question about Du Pont's experience and its relevance to the President's Commission on Automation, let me tell you, broadly, what has happened in the Du Pont Company, particularly with respect to employment. After the Korean war our total employment remained essentially constant for about eight years. What were we doing during those years? Among other things, we were spending a lot of money modernizing plants that had not been kept up to date during the war period, and we were introducing, on a very broad scale, a brand new technology that used less labor per unit. We did not increase our employment as much as we would have, if we had automated to a smaller extent and still kept all our markets. But on the other hand, if we had not automated at all, we would have had to dissolve some of our businesses, and, perhaps our total employment would have fallen.

Today, this process of modernizing is proceeding less rapidly and makes up a smaller percentage of our total construction expenditures. But within the last few years our total employment has gone up. These last few years have been very prosperous ones and the increases in our employment are just too large to be anything else but a response to the necessity of having more employment as total production goes up.

The Function of Research in a Corporation or Industry

by E. R. PIORE

Vice-President,
International Business Machines Corporation

THE FIRST THING I would like to say is that a scientist, or for that matter any technical person, is never conscious of scientific method. If you ask a person in a laboratory what his scientific method is, he will think that you are asking an irrelevant question. He works in a certain way, possesses a certain kind of wisdom and has a great body of information that has continual validity for his work; scientific methods are only a very artificial structure superimposed on this basic orientation.

It is also important to note that a physicist and a biologist, for example, approach their respective problems quite differently. This, of course, presents a problem when one tries to describe science or technology to a group that has almost no acquaintance with the daily work of a scientist. Given this basic fact, let me say that there are three fundamental types of technical activity in a modern corporation of any significant magnitude. These three activities involve manufacturing existing products, designing new products, and working on plans for the future. I am going to try to indicate to you that in each of these activities there is a very large component of research, and by research I mean the sort of work that can be done in an academic institution.

The distribution of the technical resources of a company among these three categories depends on many factors, which, at times, have very little to do with technology or research. The distribution of resources depends on the size of the company, the company's position in the market and how much of the market it occupies, the character of the market, and the restraints under which the company operates—for example, antitrust restraints, its patent position, and so forth. However, one should not overemphasize these two restraints since experience has shown that regulated companies, for instance, the telephone company and other public utilities, can operate with technological efficiency.

In the area of manufacturing existing products, we try to regularize and standardize the production process so that if, for example, the glass blower comes in one day with a cold, this does not upset the whole situation. Another thing we try to do, of course, is to reduce the cost of production. In a company like the International Business Machines Corporation, which rents much of its output, this involves, in particular, reducing maintenance costs for the machinery.

However, it should be realized that many industrial processes take place without really having a firm scientific base. For example, the vacuum tube—and most of us are old enough to have used vacuum tubes—had a barium-strontium oxide cathode filament. Millions upon millions of these tubes were made, and all through that period basic research in the field of barium-strontium oxides was going on. It was not a glamorous field and no one got a Nobel Prize, but it was an accepted area of research, and papers were published in the *Physical Review,* even after vacuum tubes had lost their crucial importance. The same can be said about the tungsten incandescent lamp. Even though there has been a large volume of production and the lamp has been a profitable and marketable item for decades, researchers continued for a long time to try to understand the process more completely so that the lamps could be fabricated more economically and so that their life in the consumer's home would be increased. This involved research on tungsten and on such things as vapor pressures of metals.

I want to emphasiz that this was really good basic research, according to the classical definition of research used by the Science Foundation. Of course one undertook such research only if one's products had been on the market for a sufficiently long time, if one had a large volume of production, and if one felt that the research might help reduce costs and increase profits. But even today, if you went to a science foundation with a proposal to do some sort of basic research, such as the study of the electron emission of barium-strontium oxides, I am sure that someone would give you ten thousand dollars for it. Never mind that in all probability it will not lead to any new concepts in physics. Most basic research does not!

Research on new products is undertaken by business firms simply because they are business ventures and because they know that new products will have to be made and will have to be improved. For example, IBM knows very well that the performance of the computers will depend on the logic circuitry and the memories in these computers. We also believe that these memory units will be magnetic in character for some time to come, but we are very uncomfortable because we do not know all the best materials that can be used for memories. Therefore IBM undertakes basic research on the nature of magnetic materials and of other materials that may become useful in making memories. We do this sort of work knowing that if we do not do it someone else will, who will get ahead of us, either by making less expensive products or by excluding us through a patent position.

We come to the last item: tentative and uncommitted activities for the future. In our present expanding economy, every company under the sun wants to get into new markets. Most companies intent on growing want to buy up other companies, but not every company has the privilege of doing this. They cannot go and find a promising young man who has just developed some promising new process in a garage, pay him a million dollars, and take over his business. They might, indeed, find themselves in jail if they tried—as a result of an anti-trust action. So, a large company may find that it has to develop new products for new markets internally, and with its own resources.

Yet even here one can find certain obstacles. One of the most difficult to overcome is dependence upon government purchases. If we examine the efforts of the aircraft industry or of any large defense industry to remove themselves from a dependency upon government purchases and to enter the commercial market, we will find that almost all the efforts fail. Such defense oriented industries just do not know how to approach the civilian market. Most of these industries are perfectly hopeless in terms of achieving profits through research. One can invest in research for technological ends, and perhaps even earn a Nobel Prize in the process, but this will not really bring in any money unless one also knows how to deal with the market.

DISCUSSION

THE ALLOCATION OF RESEARCH FUNDS: CRITERIA AND TECHNIQUES

Question: How does IBM decide where to allocate its money for research purposes?

Mr. Piore: First of all I should emphasize that there is not much free money available for research gambles, not as much as one would like to see. The whole process of research allocation is designed to protect our present inventory and our present line of products and to make sure that our future products will dominate the market. In this sense allocation is dominated by profit considerations.

In terms of the actual allocation of funds, IBM, like any big organization, is set up so that there are balances and counterbalances. If one department wants everything, there are forces that come into play which make it necessary for it really to fight for funds. It is a procedure which is quite similar to the relationship between the executive, the legislative, and the judicial branches of the federal government: The modern corporation sets up precisely the same sort of conditions in which a consensus must be achieved before any action is taken, so that no profound error can be made as a result of any particular decision. Furthermore, if a gamble is to be made, then the decision must be made at a high level.

Question: Do you think that the classifications "pure research" and "applied research" have any useful application in the industrial world?

Mr. Piore: They do not in industry. They were set up to clarify some problems in Washington.

Question: My experience in directing research in another large, technologically advanced corporation has led me to the same conclusion, and perhaps this is, therefore, of central importance. These classifications are of no use to us and we waste an awful lot of time sorting out research activities merely for accounting purposes.

Mr. Piore: More than that, one becomes involved with problems of economizing.

Question: You mentioned several types of research—research oriented toward existing products, toward future products, and so on. How do you decide on the shape of your overall research program? How much freedom of movement do you have in shaping your research program. How do you evaluate the effectiveness of your research operations?

Mr. Piore: We decide upon the general direction of our research program first of all on the basis of history and precedent. We do not start every year with a fresh view of the world. Our research efforts are shaped by the basic technology of the business that we are in. For example, at IBM, we are not doing a great deal of research involving thermonuclear energy, since we are interested in small energies. In line with this, perhaps, we should also be studying biological systems.

Given the fact that we deal with certain fundamental mechanisms involving logic and the storage of data, we try to organize a research program that basically deals with these general areas. We work with semiconductors, with magnetic resonances, with anything that deals with small energy levels or with anything that is interesting because it involves very sharp changes in state. This is also basically how we decide which people to hire.

Question: Would you say, then, that your research program is to a large extent determined by the kind of people you hire?

Mr. Piore: I think that would be correct, and I do not see how you can get away from that fact.

Question: Mr. Piore has pointed out that the allocation of resources for research depends upon a number of factors, largely nontechnological, and rather economic in character. But he has also pointed out that innovation and invention derive from technological and scientific factors, rather than from economic factors. Does this mean that research activity in the modern corporation is oriented toward maximizing profit rather than toward maximizing innovation, even though it is true that the firm is trying to create new products and new markets?

Mr. Piore: First of all I should emphasize that a research organization, which is given the function of maximizing profits for the corporation is not going to last very long as a research organization. In a corporate structure, you cannot put the responsibility of maximizing profits on the researcher's back.

But on the other hand, this leads to a very difficult problem. Profit is a very important thing. In one sense it molds the soul of the corporation. Where does one put the responsibility for maximizing profit? One certainly cannot put it on the sales division or on the engineering division. The sales division only wants to sell more and more, because that is the nature of sales activity. If one puts the responsibility on the engineering division one cannot be sure that the best possible products will be manufactured.

Question: In the final analysis, it seems to me that you still have not answered the very important question: How do you evaluate on-going research? I assume that you always have to make some hard choices and that it is on occasion necessary to turn down potentially good ideas.

Mr. Piore: This is necessarily a question of judgment. Our judgment is largely determined by what we think the needs of the company may be in the future, and, therefore, may have a strong subjective element.

THE RESEARCH FUNCTION AND TOP LEVELS OF MANAGEMENT

Question: How do your colleagues at the higher levels of management view the function of research since they are amateurs in this area and you are the professional?

Mr. Piore: They are not very comfortable with research, but on the other hand they feel that it is necessary.

Question: When a crucial decision is to be made, either to take or not to take a research gamble, to what extent do you function as a scientist and to what extent as a manager? Do you have certain insights because you are a scientist that are crucial in making such decisions, or is your competence as a scientist not as relevant as your competence as a manager?

Mr. Piore: I think that the problem you have raised has global dimensions. It is almost impossible to isolate science and technology from everything else that is going on in our society. This fusion, in fact, is most apparent in government, where you find that when the National Security Council meets a number of highly trained technical people are present. I think that this is due to the nature of our world. Those of us who are technically trained feel that all people responsible for making decisions ought to be exposed to science.

Question: In the formulation of national policy it has been argued that scientists per se do not make their decision as scientists. They make their decisions. . . .

Mr. Piore: . . . as total men.

Question: Exactly. I think we are trying to understand just what it is that a scientist contributes to the decision-making process, and whether his contribution should be more, or less, or different in some way from what it now is.

Mr. Piore: Let me refer to England for one important instance of the relationship between policy and science. Sir Solly Zuckerman is in all probability the top technical person in the British Government in the area of technological policy and defense policy. He has served in this capacity both in the Conservative and in the Labor governments. A conversation with Sir Solly indicates that he is continually trying to understand the relationship between technical facts and defense policy. But he does not say to the policy makers, "These are the technical facts; this is the policy." He tries to do much more. He interprets the technical possibilities in terms of the policy choices. He may, it is true, be biased, but this is a risk that must be run. It is no longer possible to say, "You are the specialist in science and I am the specialist in policy-making." There must be a coupling of these different roles.

THE COMPETITIVE ASPECTS OF RESEARCH AND DEVELOPMENT:
THE PROBLEM OF ANTITRUST POLICY

Question: You indicated in your talk a concern with antitrust actions. To what extent is your research activity prompted by an awareness of what is being done by other companies?

Mr. Piore: In talking about antitrust actions I was speaking in general terms, not specifically about IBM. But let us take a very specific example. IBM is deeply involved in research with semiconductors. Yet if we look at the number of people employed in research in this area by various other companies we find that IBM is nearly in the middle of the pack, in terms of the total number of people so employed. So in this area of technology we obviously do not have a dominant position. However, we want to make sure that the most advanced technology on semiconductors is applied to our products, and that we do not find ourselves in an impossible position with respect to patents. We need to be able to deal with everyone above us and below us, on cross-licenses if necessary. Technologically, there is no such thing as a monopoly. We just could not hire all the people necessary to make a technological monopoly possible.

Question: Then what is your concern with the antitrust laws?

Mr. Piore: Very large companies cannot purchase other companies in the same field. Du Pont cannot buy a chemical company; the government will not permit it. Similarly, IBM cannot buy a computer company, whether it is a million dollar company or a hundred million dollar company. But this has nothing to do with the question of domination over a particular branch of technology.

Question: Do you think this is bad or not?

Mr. Piore: Neither. What I am saying is that because of the existence of antitrust laws we have a research program that is different from that of a company which does not operate under these constraints.

Question: There has been periodic discussion of the desirability to revise the patent laws. Since you are one of the key scientists at IBM, I wonder what position you take on this.

Mr. Piore: The use of patents and their importance varies tremendously from industry to industry. In the electronics industry, for example, most of the large companies cross-license their patents. I doubt that this is the situation in the chemical industry. How important patents are, therefore, has more to do with the character of the industry than with the character of the patent laws.

Question: To what extent is your research program influenced by the existence and character of the patent laws and their implications for profits or lack of profits, as the case may be?

Mr. Piore: We are influenced a great deal and would substantially modify our research program in order to get a patent position. At the same time I would not say that we alter our research program in order to obtain profits from patents. One does not make money on a patent; one protects oneself. A large company is more concerned about protecting itself from the small, itinerant inventors than about making money on a patent.

The itinerant inventor, incidentally, is a very knowledgable person technically; he is very often associated with technical laboratories and does not take out patents in his own right. Perhaps I should add that a large company makes money by producing things, and not by having a patent monopoly.

Question: What are the potentialities for pushing the technological fronts ahead in those areas where either capital or management initiative and interest is lacking? For example, in an industry such as the railroads do you see any room for government as an instrument to bring about technological and economic development.

Mr. Piore: The government is seriously studying a proposal to build a monorail from Boston to Washington that would go at very high speeds. I do not know whether anyone has the courage to take this kind of gamble, which would probably involve an investment of around one billion dollars. On the other hand, with a problem like water pollution in our rivers, it is possible to take small steps and to watch the effects. In such areas I think that the government can make a tremendous contribution.

Question: In his remarks to the seminar last year, Assistant Secretary of Commerce J. Herbert Hollomon has urged that whole new problems related to urbanization, transportation, and so forth could be solved if they were invaded—and he used that term deliberately—by large, strong companies like IBM who would define for themselves the areas in which they wished to work.[1] You have emphasized that one of the key interests of large corporations today is to create a new demand for its products and new markets for new products. Do you see any role that IBM, for example, could play in acquainting the government and other interested parties with problems that might be solved by such an "invasion"? Could you explain to governmental policy makers the technical opportunities and the policy implications of such "invasions," and perhaps even persuade the government to take action in certain cases?

Mr. Piore: A large corporation likes to probe and develop markets by using its own resources. There are very few large corporations, except perhaps the airplane manufacturers, who would be willing to explore and develop a market partially on government resources. It would be very difficult for me to understand why a corporation should undertake an "invasion" of the urbanization problem if it knew that the government would be making part of the investment and would, therefore, be calling the tune. Such an "invasion" would have to be made purely as a public service. Private companies are very loathe to undertake such ventures because they do not know how much freedom of maneuver they will have.

Question: But the government would, in a sense, only be the customer here.

Mr. Piore: Oh, no! It would be much more than that and, in any case, the government is a supermonopolistic customer. It decides what allowable costs are; it objects to what benefits you give the workers. It even goes as far as to say, "This man should not have a rug in his office." And people just do not like to deal with monopolistic customers like that.

Question: But we built a whole airplane industry and missile

---

[1] See *The Impact of Science on Technology,* ed. by Aaron Warner, Dean Morse, Alfred S. Eichner (New York, Columbia University Press, 1965), pp. 137-40.

industry with government participation. Why couldn't we do the same thing in other areas?

Mr. Piore: You have chosen a highly specialized example, which has peculiar economic characteristics. The airplane industry has comparatively little capital invested in plant and equipment; most of it is government owned.

Question: This would be the case with the proposed monorail system.

Mr. Piore: Perhaps in that case the airplane industry might be willing to go into that area, but you asked me about American industry in general, and it is my opinion that companies such as U.S. Steel, Du Pont, and IBM would want to think through very carefully any such undertaking before taking part in it.

INNOVATION AND MARKET RESEARCH

Mr. Piore: I would like to make a comment about the relationship between invention and innovation and industry's approach to producing and marketing a new product. We have reached a stage in our economic development where we put tremendous emphasis on operations research and upon market analysis—on the whole formal procedure, in a word, of assuring management that the gamble of introducing a new product will be at a minimum risk. But I have yet to see one case in which there has been a profound growth of a company by means of the procedures which are being taught in the nation's business schools these days. One can find very thoughtful operations research people, very thoughtful market analysts who carry out very careful studies of such things as population and age distribution, but, it seems to me, that when real growth occurs it is because some person in the laboratory, some technical person usually, senses what the developing needs of the economy are, and some man in the management of the concern has the courage to probe the market. Only then does the product begin to sell.

Question: I would have thought that it was appropriate to use these techniques of market analysis to start a technological breakthrough of the sort to which you referred in your talk.

Mr. Piore: First of all, enthusiasts of operations research claim that they have the answers to everything once they have put a problem into a computorial shape and described it with a few probability curves. This claim is simply not correct. Secondly, to be truly effective the operations research specialists have to report to the highest place in management and it is difficult to get immediate action on that level. The basic problem, of course, is that they think that they have all the answers, and I am simply claiming that life is a little more complicated than that. Computers, like other instruments, ought to be used only in the right place. Operations research cannot be a cure-all for unimaginative and unenterprising management.

Question: What determines the point at which you stop trying to protect your existing inventory and move on to the next step, the new product? What determines the point at which you shift research resources from what you define as a new product, which will definitely be produced, to the future activities to which you are as yet uncommitted?

Mr. Piore: This question can be answered very simply. The shift of resources is determined by one's assessment of the market, of the nature of the competition, and of the requirements for survival. If you were to ask what factors go into these assessments, that would be a long story, which could not be described with just simple formulae.

Question: To what extent do you create markets, rather than wait for them to evolve naturally?

Mr. Piore: The creation of markets leads to the most desirable type of shift in a firm's activities, but it is one which is very difficult to achieve. Every firm wants to create new markets and most large companies have a department devoted exclusively to that problem.

Question: It is generally believed today that we are going

through a period of enormous acceleration in the rate of technological change. Could you indicate, by means of the interplay
between your three categories of research, the factors which
determine the speed with which a process undergoes change?
For example, you mentioned that computer memories would
continue to be magnetic as far ahead as you can see. Could you
tell us what factors might cause that situation to change, and
what effects this would have on the three categories of research
on magnetic memories?

Mr. Piore: I will confine my answer to the question of magnetic memories. The best computers are going to become
smaller and they are going to operate faster. This will have
certain consequences in the area of production. The application
of whatever innovations are made will involve a smaller amount
of human work, simply because of the character of the new
technology. You just cannot use human labor; the human eye
and hand are simply not fine enough, not steady enough. Sometimes automated production is the only way to make a component.

Another implication of the faster rate of technological change
is the growth in the size of research facilities. If we observe
the development of laboratories over the last ten years, we
notice that they have required more and more square feet per
person, because a researcher needs more and more instruments.
There is very little science left that can be done just by using
sealing wax and string. We can no longer depend on having
an acre of draftsmen counting out parts or setting up experiments.

Question: Are the marketing, selling, advertising and pro-
motional costs of a company like IBM going up or down?

Mr. Piore: In the case of IBM, marketing costs have gone up
as engineering costs have increased because, when we talk
about marketing at IBM, we must include the servicing of the
customer. We have to think of the customer's computer program and make sure that he is not wasting machine time. So we

have systems engineers who go to our customers and say, "You are not doing things right; you have got to reprogram," and so forth. This is how we do business, and this very expensive service is all charged to marketing. Therefore, I could not give you an estimate of our marketing costs in conventional terms.

Question: Do your responsibilities as Senior Scientist at IBM include the area of social science research? Could you comment on this aspect of your work?

Mr. Piore: My responsibilities do include the supervision of social science research, but I am not sure that my understanding of the meaning of the term, social science research, would be the same as yours. Let me list what I think we do in the area of social science research. We have a staff of economists; we also have a very large educational operation; we also support the research of some mathematical economists who work on problems such as the optimum use of inventory; finally, we have some people doing research on what one would call human factors—such things as making sure that tables are at the right height and so forth.

Actually, to my mind, the people who are dealing most directly with the revolution in the social sciences are those who are trying to develop what might be called generalized languages, which enable a person to use a typewriter and operate a computer directly, without any complicated programming. There are a lot of people in the engineering division who are engaged in this type of research.

THE INVENTOR AND THE RESEARCHER IN INDUSTRY

Question: Could you say a few more words about invention, since the informal title of your talk was "How to Turn Technology into Profits"?

Mr. Piore: It is very difficult for most people to view the concept of invention realistically. There is a popular notion that an inventor is a man who has a lucky idea. I think that this is just not true historically. A study of the history of science indicates that those inventors who have had a significant impact on society were people who knew science very thoroughly. For example, the Wright brothers really tried to understand the flow of air over a wing and the principle of lift, because all the information they had on these subjects at the time was wrong. Similarly, most people think of Watt as just another itinerant engineer, whereas in reality he had a great deal of scientific knowledge. In my own experience some of the best inventors I know are people who are among the world's most sophisticated experimental physicists.

Question: We in the academic world frequently hear complaints from industry that the kind of research man we prepare for you is not really satisfactory. Is this true? And could you say something about the kind of freedom of research which a gifted research man has at a company like yours.

Mr. Piore: There is a perpetual argument going on at IBM over the question of research freedom. Even in the academic world a man has to earn the right to have freedom of research. In the academic world he is put through a very tough grind— through a doctoral program and postdoctoral work—before he gets that freedom. We at IBM have the same tough standards. Merely because a man has a Ph.D. does not, as I see it, give him the right to freedom of research. He is a young man, still an apprentice in many ways. There are, of course, exceptions, but these are rare.

As for the general caliber of researchers turned out by the universities, I have no complaints. If they are from good universities, there is no problem.

Question: Could you give us a notion of the extent to which you engage in a systematic effort to scan the literature and to review the results of the research of other institutions and research centers. And do you consider applying these research results to production and research in your own company?

Mr. Piore: When we give a person at IBM technological responsibility for an area, we expect him to know everything about the literature and the patents in that area. Our vice-presidents try to organize this kind of material and see to it that it is distributed to everyone. The problem is to get people to read it.

Your question is symptomatic of a common attitude. It is almost impossible to convince a layman that one of the hall-marks of a good technical person is that he knows what is going on in his field, seemingly by a kind of intellectual osmosis. He has kept up with developments all his life without having the information organized for him. Today we are trying to organize and structure this process, but I am not sure what good it will do. The Russians have tried to do just this, and I haven't been able to see what good it has done them, except that they know Western literature better than most Westerners. It hasn't produced better science or better products in their case, as far as I can see.

I must add that it is amazingly hard to suppress an important technical accomplishment. Even in a competitive industry somehow it always becomes diffused.

PART TWO

The Social Impact of Technological Change

Introduction

THE TONE of the papers in Part One is one of cautious optimism, and the approach to problems is particularistic and experimental. The scientists, at least those represented here, seem to indicate that social change (and along with it, social progress) takes place in piecemeal fashion; problems can be isolated and solutions can be worked out by rational techniques. The papers in Part Two have a very different tone. Here the pervasive atmosphere is one of apprehension, not local but general. Problems are seen in larger terms, invading every area of existence, and although the rational mind may discern solutions, the application of such solutions may often be blocked by ignorance and irrationality.

Perhaps this difference in tone stems, in part, from the fact that the authors of these papers are either social scientists by profession or are in some other way concerned with the behavior of very large numbers of people and in their work must take into account social and political attitudes which frequently seem to be based largely on ignorance and unreason. Two of the authors represent major branches of the social sciences. Henry H. Villard is an economist whose professional interests have centered around problems of economic development, and Donald N. Michael is a sociologist and social psychologist whose work has often dealt with the social and psychological concomitants of technological change. The work of William H. Sebrell, Jr., Director of the Institute of Nutrition Sciences of Columbia University, has brought him into contact with a wide range of peoples at different levels of social, political, and economic development. And Joseph S. Clark, a lawyer by training, has been Mayor of Philadelphia, and as United States Senator from Pennsylvania was chairman of the Subcommittee on Manpower and Employment of the Committee on Labor and Public Welfare, which recently held extensive hearings on the general problems raised by technological change and automation as they bear upon the nation's labor force.

Part Two opens with Dr. Sebrell's paper which asks the challenging question, "Can we feed the world?" The discrepancy between what is technologically feasible and what is culturally and economically possible emerges as one of the central points of the paper. Pointing out the vast potentialities for expansion of conventional food production and the even greater potentialities which lie in new sources of food supplies and new types of foods, Dr. Sebrell at the same time gives specific instances of intransigent opposition to the effective utilization of these remarkable recent developments in food technology.

It is not enough, insists Dr. Sebrell, for the scientist engaged in the development of novel sources of proteins and carbohydrates to limit his investigations to the scientific requirements for adequate nutritional levels. Perhaps even more essential is a thorough awareness on the scientist's part of the extremely subtle and complex character of existing food habits and, along with this awareness, the ability to find techniques to break down resistances to dietary changes. Dr. Sebrell is impressed by the success with which commercial food manufacturers and distributors surmount such problems, and suggests that in order for the technological opportunities of modern food production to be satisfactorily exploited there must be close cooperation between the scientist engaged in nutritional work and the commercial organizations who know now how to exploit market opportunities.

Yet it would be fair to say that an even more general point lies at the heart of Dr. Sebrell's paper, a point that finds echoing responses in other papers in this book. The scientist who engages in nutritional studies cannot work in isolation. A very careful coordination, combining a historical sense on the one hand and a sense of the emerging problems and needs on the other, is essential to success.

It is significant, indeed, that all of the papers in Part Two contain projections of future needs of one kind or another and advocate policies designed to cope with problems whose boundaries are just beginning to become perceptible. The authors share the belief that the kind of problems they are concerned with are more than local and ephemeral. But the time periods that they are concerned with are quite varied. For example, Mr. Villard asks the reader to consider the world's population in the year 2000, and he even speculates briefly about projections covering much longer periods. Similarly, Mr. Michael is concerned with a period of time that extends well into the future. But Mr. Michael emphasizes, in particular, that many of the problems that have been brought into existence by technological change and that are now becoming critical are the result of technologies of the past. He asks the reader to see such problems in their full historical framework.

Both Dr. Sebrell and Senator Clark are strongly impressed by the immediate urgency of the problems they are concerned about. The immediate tasks are so enormous that both men feel that it is almost a luxury to be able to cast one's thoughts into a future that extends more than a few decades ahead of us. And both authors betray some suspicion of the value of very long-range projections to the man of affairs—to the politician or the nutrition expert.

There is, however, a striking difference between the tone of Dr. Sebrell's remarks and those of Senator Clark. Dr. Sebrell shares the pessimism, which is such a marked characteristic of both Mr. Michael's and Mr. Villard's paper. After detailing some of the promising developments in food technology, which might lull the reader into a state of confidence, Dr. Sebrell brings his paper to a sharp conclusion by reminding the reader that the long-range solution of the problem of world food supply depends, in the last analysis, upon man's ability to control his rate of reproduction and he points out that "there is little evidence at the present time to indicate that control of reproduction will become widespread enough to prevent a disaster."

Senator Clark, on the contrary, although admitting the complexity and intransigency of particular political and social problems, has high hopes that America's ingenuity will be stimulated by the very immensity of the challenges to her resources. He presents a compelling argument that the economy must grow at exceptionally rapid rates in the decades immediately ahead if it is to absorb the unprecedented number of new entrants into the labor force and at the same time make substantial inroads upon current levels of unemployment. At the same time, Senator Clark points out that the total amount of public and private investment required to meet pressing national needs over these same decades is so great that the nation's resources of men, capital, and raw materials may, in fact, not be sufficient to carry out investment projects that are more urgently needed with each passing year.

Senator Clark outlines some of the legislative and administrative obstacles in the way of rapid and adequate government action in many of these areas of national need. But he believes that when the urgency of a national problem reaches a certain level, it becomes possible for the nation to attack such problems with the vigor and scope that the situation demands.

Senator Clark limits his analysis primarily to the United

States. Mr. Henry Villard, on the other hand, points out that the demands that will be made on the resources of the country in the next decades must be viewed in the light of world needs, rather than national needs. Because of the staggering size of these needs, which spring in large part, but not entirely, from the rapid rates of increase of population in the underdeveloped regions of the world, Mr. Villard is convinced that the United States economy must grow at very rapid rates in the near future. In direct contradiction to those who maintain that technological change will produce increasingly severe problems of unemployment, Mr. Villard insists that the hope of the world lies in the possibility that technological progress will proceed rapidly enough in the next few decades to provide the increase in output essential to maintain standards of living throughout the world above the disaster level. A failure to do so will, Mr. Villard believes, produce intolerable political strains and increase enormously the danger of a military conflagration.

Although Senator Clark, Dr. Sebrell, and Mr. Villard are all concerned with very large problems whose solution entails a sensitivity to political and social factors, they still focus their attention individually on particular aspects of the problems related to technological change. Mr. Michael asks the reader to consider quite another and wider set of problems that he believes are associated with technological change.

He emphasizes that technological change in the recent past, and even more clearly in the next few decades, has unique features when considered against the background of technological change in the past. In the first place the rate of technological change is unprecedentedly rapid. But more important, the character of technological change today raises simultaneously more serious problems of adjustment in a larger number of areas of man's existence than it has in the past.

Mr. Michael is very much aware that the same technological progress that has created some of these problems has also brought about immense benefits, and promises even greater benefits in the future. The beneficial results, says Mr. Michael, can be expected by and large to take care of themselves. But in his view we are ill-equipped to handle the range and severity of the social and psychological problems that are being created by contemporary technology and threaten to increase markedly in the future.

The social sciences in particular, according to Mr. Michael, find it difficult, if not impossible at times, to cope with these emerging problems, because they are apt to cut across conventional academic disciplines. As a result they are the province of no single discipline and are either neglected entirely or at best given only perfunctory attention.

Finally, Mr. Michael emphasizes that the problem of employment, frequently the center of attention of those who are worried about automation, is not really the subject of his concern except peripherally. Rather, he is concerned whether we will be caught unawares, and, therefore, unprepared, by the tremendous strains which sophisticated technologies in many fields, including the biological, will place upon our traditional morality, upon our traditional political processes, and most significantly, upon our very conception of the nature of the individual and his relation to society.

The theme, then, of challenge and opportunity, which pervaded the papers of Part One, unifies in a like fashion the papers of Part Two, but here the challenges are often seen to be more threatening and intractable. Bringing population increases under control in the very limited number of years that may be left to us, and feeding the very large increases in population that will in any case take place is sure to tax our scientific and technological ingenuity. Providing the peoples of the world with satisfactory standards of living cannot but impose immense burdens upon our production capacity. These burdens can be alleviated, to some extent, by scientific discovery and technological progress, but it will require much creativity on the part of the social scientist and the physical scientist working together, if societies in various states of social and economic development are to make satisfactory use of what will be technologically possible. The papers of Part Two remind us that scientific discovery and technological advance can be fruitless —indeed, in certain cases may prove dangerously disruptive and corroding—if the politician, the social scientist, and the philosopher do not participate actively and constructively in helping to solve the problems created by the impact of technological change on society.

Can We Feed the World?

by WILLIAM H. SEBRELL, JR.
*Robert R. Williams Professor of Nutrition,
Columbia University*

IT IS A GREAT PLEASURE for me to have this opportunity to stir up some discussion with people whose disciplines are different from those of my usual associates. Before I get to my topic, let us clarify two or three basic concepts that we must understand to make my discussion more meaningful.

First of all, the human body must, above everything else, have a supply of energy. This we measure in calories, which are energy units. The body wastes away and dies very quickly if there is not enough energy to keep the heart beating and the lungs moving. Hunger pains occur and a person wastes away so rapidly that he becomes desperate and will go to any lengths to obtain food, will even kill somebody and, possibly, even eat another person. Under those conditions, what man will eat is determined by his terrible drive for energy, and he will eat literally anything. This has been proven time after time by serious famine situations. Man gets energy primarily from two types of food—carbohydrates and fats, which are the basis of the food supply of mankind.

There is another type of food that is equally important for survival, and this is protein. The body tissues—that is, the muscles, and organs—are made up of material that contains protein. This protein has to be synthesized from compounds called amino acids. There are eight amino acids that the human body cannot manufacture, so they must be obtained from food and so must an adequate amount of crude nitrogen. If any one of them is missing, the body cannot make muscles, cannot function effectively, and degenerates.

A lack of protein is most serious in a young child, because the young child is building muscles and other tissues very rapidly from birth until it is six or seven years old. If a child does not get an adequate supply of protein, it will die. An adult becomes sick, cannot work, and wastes away, but he does not have hunger pains. Thus, a population will live on an inadequate supply of protein and will do so without protest, because it does not know what is wrong. So, in looking at the question of whether the world will have enough to eat, we must first look at carbohydrates and fats, and, secondly, at protein and the essential amino acids.

If an individual does not have enough calories in terms of carbohydrates, his body will use protein to supply the needed calories. Therefore, we cannot separate the problem of calories and protein. They are both connected with a condition known as calorie-protein malnutrition. In other words, to make the most effective use of protein and amino acids, a person must have enough calories from carbohydrates and fats. Therefore, we must consider the problem of feeding the world in terms of foods that supply calories, protein and the essential amino acids.

In looking at it this way, let us first mention the question of population. The estimated population of the developed areas of the world in 1960 was 857 million. The projections to the year 2000 indicate that the population of the developed areas at that time will be 1266 million. In the underdeveloped areas, the population for 1960 was estimated at 2153 million and the projection for the year 2000 is 4699 million. Assuming that the projections are correct, the greatest increase in population is going to occur where the food shortage is greatest at the present time. In addition, if we are going to provide adequate diets for everybody in the year 2000, we must increase food production by about 200 percent.

It is not enough to talk about food production. The food must also be moved to and consumed by the people that need it most, and this is one of the cruxes of the whole problem. It is both a problem of transportation and production; I will return to this matter a little later.

A question that is often asked is: Will a shortage of food so starve the population that a wide-scale famine will readjust the population? It is one possibility, certainly. But I think I can demonstrate that, if we apply the technology we have, it is not going to be food that will be the ultimate factor in limiting populations.

That famine will strike the developing areas of the world is almost certain, because food production in those areas is not keeping pace with population growth and cannot be accelerated fast enough. However, the famine will not be great enough to limit the population directly, although its indirect effects may help accelerate population control.

One of the reasons why we have a population explosion today is the advance we have made in public health and in the control of infectious diseases. If we look at health figures, we find that we are preventing deaths mostly in the younger age groups; we are not really doing much to prolong the life expectancy of old people. The reason for this is that we now have the means of controlling infectious diseases by immunization, antibiotics, and sanitation, so that children today do not die of measles, pneumonia, smallpox, and all the other epidemic and infectious diseases as they used to in the past. It is this decreased infant and child mortality that is causing the population explosion.

Malnutrition is one of the most important remaining causes of death among children. It is estimated that more than 3,000,-

000 children die annually from malnutrition. In some countries 50 percent of the babies die between the ages of two and six, largely because of malnutrition. This fact is hidden in the statistics of a developing country because diarrhea, parasites, and diseases of that kind are recorded as being the cause of the children's death. But they actually die of protein-calorie malnutrition. If they were well nourished, they would not die.

Some people callously say, "Why save these children? Let them die. Don't improve the food supply. We've got too many people already." My answer to this is that no matter how we control the population growth, we should not do it by allowing babies to die. Some other solution must be found to this most urgent problem of how to control population growth—particularly in areas where the food supply is already inadequate.

TABLE 1

	Period	*Calories available per person*[a]	*Percent of requirement*	*Protein available per person*[b]	*Percent of requirement*
Low-calorie countries	Prewar	2110	91	62	103
	Postwar	1960	84	56	93
	Recent	2150	93	58	97
High-calorie countries	Prewar	2950	115	85	142
	Postwar	2860	111	85	142
	Recent	3050	116	90	150

Source: Prepared from figures of the Food and Agriculture Organization of the U.N., in *Third World Food Survey*, Basic Studies, No. 11 (Rome, 1963).

[a] Total daily calorie requirement per person for low-calorie countries is calculated to be 2320, for high-calorie countries 2580.

[b] In grams, based on "standard" 60 grams daily total protein requirement.

What is the present food situation as compared with the past? Are we improving the world food supply or are we losing ground? I want to call your attention to Table 1. Here the world is divided into two regions, low-calorie countries and high-calorie countries, and three periods of time are shown, prewar, postwar, and recent. The next column shows the calories available per person, and the column after that gives the percentage this is of the actual calorie requirement. The low-calorie countries met 91 percent of their calorie requirement in the prewar period, but in the postwar period they met only 84 percent of their requirements. Now the figure has gone back to 93 percent. In the high-calorie countries the figure has remained well above the minimum requirement. The protein available is in the next column, and this is given without regard to the quality of the protein. The protein requirement is set at 60 grams. You will notice that the low-calorie countries met their requirement before the war; they dropped to 93 percent of requirements during the war. Recently they have been able to meet only 97 percent of their requirements. The figures for the high-calorie countries are higher than prewar levels.

What has happened is that the population has increased faster than the food production, and the result is that the low-calorie countries have never returned to even their prewar level of per capita food intake.

In Table 2, this same kind of data is shown for these areas, Western Europe, Eastern Europe together with the USSR, and North America. You will notice that all of the countries in these areas are above the 100 percent requirement both in protein and calories. In other words, these parts of the world, on this rather crude basis, do not have any problem. But now look at Table 3. Here some countries in South America are compared with some countries in Asia and with Indonesia, and the Philippines. Note the low level of available calories. And the percent of their protein requirements shows the same characteristics. The United Nations statistics show that some of the countries are actually continuing to lose ground: their population is growing faster than the increase in their food supply.

While these statistics look quite serious, we have still not considered another basic problem, which is the actual distribution of the food to the individual. Well-to-do people in these

TABLE 2

	Calories available per person[a]	Percent of requirement	Protein available per person[b]	Percent of requirement
Western Europe				
Prewar	2880	112	85	141
Postwar	2750	107	82	137
Recent	2910	113	83	138
Eastern Europe and USSR				
Prewar	2850	110	84	140
Postwar	2780	108	82	137
Recent	3180	122	94	157
North America				
Prewar	3260	125	86	143
Postwar	3170	122	91	151
Recent	3110	120	93	155

Source: Prepared from figures of the Food and Agriculture Organization of the U.N., in *Third World Food Survey*, Basic Studies, No. 11 (Rome, 1963).

[a] Total daily calorie requirement per person for Western Europe is calculated to have been 2570 for the prewar period and 2580 for the two other periods; for Eastern Europe and USSR, 2600 for all three periods; for North America, 2590 for all three periods.

[b] In grams, based on "standard" 60 grams daily total protein requirement.

areas do not have any trouble getting food. It is the poorer classes that are not getting the food, so that the nutritional deficit is concentrated in only one part of the population, which means that it is much more serious than the figures show.

What is required in the near future? There appear to be some deceptively simple answers: Put more land into cultivation, increase the yield from the land that is being cultivated by fertilizing it better and by using better seed, make better use of what is available, catch more fish. All of this sounds very good, but the problem is much more complicated. First of all, only about 10 percent of the earth's surface is being cultivated. About 71 percent of it is covered by the oceans. A fair amount of it is mountain, desert, and jungle. It is easier to say, "Put more land into cultivation," than to do it. The jungle areas are nonproductive; cleared jungle land is not fertile because the heavy rainfall has leeched the nutrients from the soil. The

William H. Sebrell, Jr.

TABLE 3

	Calories available per person[a]	Percent of requirement	Protein available per person[b]	Percent of requirement
Bolivia, Chile, Colombia, Ecuador, Peru, Venezuela				
Prewar	1970	78	55	92
Postwar	2160	87	53	88
Recent	2190	88	56	93
Ceylon, India, Pakistan				
Prewar	1950	85	52	87
Postwar	1720	75	46	77
Recent	1970	86	50	83
Indonesia, Philippines				
Prewar	2020	89	46	77
Postwar	1900	84	42	70
Recent	2070	91	45	75

Source: Prepared from figures of the Food and Agriculture Organization of the U.N., in *Third World Food Survey*, Basic Studies, No. 11 (Rome, 1963).

[a] Total daily calories requirement per person for Bolivia, etc., is calculated to have been 2510 for the prewar period and 2480 for the two other periods; for Ceylon, etc., 2300 for all three periods; for Indonesia and the Philippines, 2270 for all three periods.

[b] In grams, based on "standard" 60 grams daily total protein requirement.

climate in the Far North is too cold; in other areas the water supply is inadequate. So there are limitations on the amount of land that can be put into cultivation.

Increasing the yield on cultivated land seems to be something that could easily be done. By the proper use of fertilizers, by selective breeding of stock, by control of pests the productivity of a given piece of land can be raised to a considerably higher level, as has been amply demonstrated in the United States. But there are certain factors that operate to prevent this. These are mainly ignorance on the one hand, and economic factors involving such things as commodity prices, farm labor, land tenure, on the other, which operate to make it uneconomic to use more fertilizer.

Why not increase our supply of food from the sea? It is true that the ocean has a tremendous number of fish in it, but there are large areas of the ocean that are just as barren as desert land. We do not know too much about fishing, but what we do know is very important. There are fertile areas in the ocean; there are areas in the ocean in which upwellings occur, and these are the fertile areas, because the current brings nutrients up from the bottom. The upwellings in the sea have not all been identified, but life in them is very prolific.

Off Peru there is an area about 200 kilometers long and about 100 kilometers wide in which this upwelling occurs. It is constant and can be relied on, and, as a result that area teems with fish. During 1964 the yield has been something like 15 million tons of anchovies from this one area. In 1954, Peru caught 50,000 tons of fish. In 1964, ten years later, the yield had increased to 9 million tons. But today Peru is one of the countries that has widespread protein-calorie malnutrition. The fish would solve this problem, but about 90 percent of the fish is exported. In fact, Peruvian fish meal has a large role in determining the world price. The fish meal is sold in many parts of the world; crude fish meal has a dark color and must be processed before it becomes satisfactory for human consumption.

One of the serious limitations of using fish in the past was the need for refrigeration, quick canning or freezing—all of these processes are expensive. Therefore, present efforts are directed largely toward making a suitable fish flour or protein concentrate. There are many technological problems in this process, but a good product can be made from whole fish. It is bacteriologically safe, there are no toxins in it, it is tasteless and odorless, and it can be eaten in bread, in crackers, and in soup. However, the Food and Drug Administration will not permit the sale of such a product in the United States, giving as reason that the American housewife would regard it as an unaesthetic and filthy product because it contains the intestines, the eyes, and all the other parts of the fish. We eat sardines, oysters, and various other sea food whole; sometimes certain parts of an animal, such as chitterlings, are eaten, and that does not seem to be unaesthetic. But at present federal regulations will not permit the sale of the whole fish product in this country. This is of some importance, because the impression is created that the product is inferior, not fit for use in the United States, and thus marketing it abroad becomes difficult.

In spite of the fact that there are no easy answers, it is quite clear that food production must be increased. The United Nations has estimated that by 1975 the world's cereal production must be increased by 35 percent, the legumes by 85 percent, and the animal products by 50 percent. This is a big order. And that does not even mean everybody will be well fed; it merely indicates the level of production needed to keep up with the estimated population increase.

The supply of calories can easily be increased. But it is not enough just to increase calories. We must also increase the supply of proteins and of the essential amino acids.

TABLE 4.   RELATIVE YIELDS OF ANIMAL AND
VEGETABLE PROTEIN

Product	*Yield* *(pounds of protein per acre)*
Grass	600
Legumes	370
Wheat	269
Milk	90
Beef	54

Source: Prepared from figures of the Food and Agriculture Organization of the U.N., in *Third World Food Survey*, Basic Studies, No. 11 (Rome, 1963).

What are the sources, then, to which we can turn to get more protein? Tables 4 and Table 5 throw some light on the problem. The yield in pounds of protein per acre given in Table 4 show that you get the most protein from grass, next from legumes, next from wheat, next from milk, and, last of all, from beef. Rice would come between wheat and milk. Wheat is from 9 to 12 percent protein, and rice is about 7 percent protein.

Table 4 also reveals how inefficient animals are in making protein. An animal is really just a factory for processing foods humans cannot eat. Animals clean out the toxins, refine low-grade protein, and make animal fat; thus they make a high-grade palatable product out of something that we cannot eat. But they do that very inefficiently. It is a mistake, however, to do away with edible animals just because they are inefficient, because, though they are inefficient, the animals eat many things humans cannot eat. Corncobs, urea, and molasses which humans would not want to eat and could not digest, can be put into a cow's diet, and will make highly nutritious beef. It should also be pointed out that the land used by food animals frequently is not suited for the production of crops acceptable for human consumption.

Table 5 gives the relative cost of protein. One pound of cottonseed flour costs 11¢. Cottonseed flour is 55 percent protein, so that on a protein base it costs 20¢ per pound of protein. Beef,

TABLE 5
RELATIVE COST OF PROTEIN FROM SELECTED RAW MATERIAL
SOURCES IN THE U. S.

		Protein	
Protein sources	Cost (in dollars per lb.)	Content percent	Cost (in dollars per lb.)
Cottonseed flour	$0.11	55	0.20
Soybean flour (food quality)	0.054	52	0.11
Fish meal (feed grade)	0.10-0.12	75-85	0.14
Wheat (Kansas City)	2.20[a]	12	0.30
Dry skim milk	0.144	35.6	0.40
Wheat flour	0.066	11	0.60
Chicken (dressed)	0.30	20	1.50
Beef (retail)	0.80	18	4.44

Source: Prepared from figures of the Food and Agriculture Organization of the U.N., in *Third World Food Survey*, Basic Studies, No. 11 (Rome, 1963).
[a] Per bushel.

however, costs $4.44 per pound of protein. This is one of the reasons why we turn to vegetable products rather than to animal products when looking for new and cheap sources of protein.

The problem with the vegetable proteins is that they do not have the right amounts of all the essential amino acids. Thus, rice and wheat flour as sole sources of protein cannot adequately sustain life and health. However, if the amino acids lysine and theonine are added to wheat flour, it becomes nearly as good as milk as a source of protein. Lysine can be made synthetically in large amounts at a price that is economically feasible.

The only practical food sources of protein we have available to us in the world are animal products, such as meat, fish, milk, and eggs; and vegetable products, such as the oil seed press cakes; and the many varieties of beans. Although the individual vegetable sources of protein do not have the right mixture of the essential amino acids, they can be mixed to achieve a satisfactory balance. Some of these mixtures have already attained wide popularity.

The oil seed press cakes, which are of the greatest interest and which are in the largest supply are those made of soybeans, cottonseed, or peanuts. There are millions of pounds of these in various parts of the world. Most of them are used as fertilizer or as animal food, but they are good sources of protein at a relatively low price. Cottonseed flour costs about 20¢ a pound of protein, whereas soybean flour costs about 11¢ a pound of protein. One mixture of particular interest is Incaparina, which is being used in Central America. This mixture will help to meet the protein needs of Guatemala and Central America very cheaply.

There are a number of other protein food mixtures in use in various parts of the world. In India a mixture is being used that is made of chick-peas, peanut flour, and dry skim milk. In Africa another mixture, made of peanut flour and dry skim milk is being used. At the Columbia University Institute of Nutrition Sciences, the staff is working on a mixture for the Middle East made of parboiled wheat, chick-peas, and skim milk powder. It is being tested on babies in Beirut, at the present time, to determine whether it is acceptable.

Another technical development is the isolation of protein from vegetables and the manufacture of products from it that taste like fish, ham, bacon, chicken, and beef. It is really remarkable how well the structure and the taste of various animal products can be simulated. The basic material may be soybeans, peanuts, and similar products.

The problem, however, is much more complex than just developing and marketing a new product. One of the problems is having the group that needs the product accept it. One revealing example is the following: An experiment to increase corn production by introducing hybrid corn was tried in the Rio Grande Valley. A group of 84 farmers in a small area was put under study. They were raising the old-fashioned strain of corn. Discussions were held with them on changing to hybrid corn in order to double their yield. A demonstration plot was sown, and it was proved that they could double their corn yield. The next year, 40 of the 84 farmers planted the new hybrid corn, and doubled their corn production. The second year, 60 of the 84 planted the hybrid corn and doubled their corn production. The third year only 30 planted it, and the fourth year only 3 out of the 84 had not gone back to the original strain of corn. What was the matter? The answer was really very simple. The hybrid corn would not make a good tortilla. The dough characteristics were such that the tortilla broke apart when the wife started to roll it.

I mention this example just to point out how important various subjective factors are in determining whether a food product is acceptable. It is not enough just to study the technological problems. We know very little about food habits; they are established very early in life, and they are very strong and very difficult to change. It is almost impossible to change established food habits unless some serious stress is involved, such as starvation or severe economic pressure.

The successful introduction of a new food in any country is a very difficult thing indeed; it involves customs, taboos, religious restrictions, and superstitions of all kinds. But the greatest problem is ignorance and illiteracy. Of course there is an important economic factor, but the basic problem is ignorance. For instance, some people do not understand that there are differences in the nutritive value of foods. They have learned from experience that certain vegetable and animal products can be eaten and that others cannot be eaten without serious consequences. They do not understand that babies must have certain kinds of food that differ from that required by adults. So when a baby is weaned he eats from the family food pot, and in many cases there *is* only a pot in which a stew or thick soup is made from all the food available. The adults do all right, but the baby gets sick. They reason that the baby's illness cannot be due to the food because they all ate out of the same pot.

Consider another situation. In the south of India there are thousands of babies who are going blind because of vitamin A deficiency. Their eyes are permanently destroyed. All that is needed to prevent this is to give the babies some green leaves. They need carotene (provitamin A) found in any green or yellow vegetable. But their mothers do not know that this would prevent the blindness, and they do know that if they gave the baby uncooked green leaves, it would probably die of diarrhea or dysentery. There is no educational means for getting this simple message to them, partly because of the illiteracy. An illiterate has no means of referring to what you have taught him.

Now I would like to turn briefly to a serious problem affecting attempts to increase local food production. It is known that if Japanese methods of rice cultivation were applied in India, and if adequate amounts of fertilizer were used, it would be possible to triple the rice production. One of the obstacles is that there is no motivation to do so. It is no use to provide the farmer with fertilizer, to give him seed, or to lecture to him, unless he is motivated to take advantage of these things. Why is he not so motivated? This question was explored in a Ford Foundation study in 1959 called "The Food Crisis in India." The answers closely resemble those that solved the problem in the southern United States after the Great Depression of 1929.

The Indian farmer lacks the motivation because he is in debt. He does not own the land and he has to borrow money at a usurious rate of interest. Any increase in food production, which frequently requires that he borrow money to finance the additional crop, is not likely to benefit him very much. Unless the farmer can be shown that an increase in crop production will be of benefit to him, all the efforts to increase production are not likely to be very successful.

Unfortunately, the enormous amount of surplus food from the United States has not been distributed or used in a manner that will stimulate local food production. Nor has it been used in the best way to combat malnutrition or to stimulate local initiative or, in many cases, the local food business. I visited a Moslem village in the Middle East where wheat was once grown; the principal cash crop is now marijuana. I asked the village headman, who was well-educated, why he was raising marijuana instead of wheat, especially since it is illegal to raise marijuana and injures the people who use it. His answer was that the wheat from the United States had depressed the local price so that his village could not raise wheat and make a satisfactory profit. He probably meant that there is more profit in marijuana than there is in wheat, but he was able to use the argument that the United States wheat was decreasing the profit that formerly could be made from producing local wheat. Complaints have been made by other countries about the adverse effect surplus foods from the United States had on the prices of locally produced foods. Such surplus food must be distributed through the regular commercial channels at a normal profit, since a permanent solution to the world food shortage requires a maximum production of food from local sources.

New food products must also be distributed through regular commercial channels. Incaparina, for example, has now been taken over by the Quaker Oats Company. Originally, it was developed in the Central American Institute for Nutrition and was a failure, but now Quaker Oats is beginning to make money out of it. It is apparently selling very well. However, we still tend to develop these mixtures and tell people they should eat them, when in my opinion what we should do is ask, "What will these people buy? How do we create a market?" And then we should try to meet that demand. Until we approach the problem in this way, I do not think we will achieve any real solutions.

In the future it is most likely that we are going to have many new products. By increasing the use of modern food technology throughout the world it is possible to make tremendous advances in the use of food. Parts of many food crops are wasted because even the simplest technological knowledge is not available to prolong the crop season, to preserve perishable foods, and to prevent loss from insects, rodents, and other pests.

Given enough time we can undoubtedly increase the world's food production enormously. However, the population is increasing so fast that it is likely that there will not be enough time to get the necessary programs in operation before the per capita food production falls below the present level. Furthermore, the areas that need the food most are the ones that are already losing ground. The big problem is not only to produce more food but to get it to the people that need it. Another complication is caused by the movement of large numbers of people from the rural areas into the big cities all over the world. This is creating large slum areas in which the food supply problem is even more involved than it is in rural areas. The population movement is caused, in part, by the great desire of the farm laborer to escape from what is often regarded as the lowest economic class, and, in part, by growing industrialization. Thus, more food must be produced by fewer people, and the marketing of the food requires more transportation and greater use of technology.

In the long run, a time must come when the population growth is balanced with the world's food supply. Biological balances are constantly maintained throughout nature, and a biological balance between man and his environment is going to come about in some way—either through some new virus, social upheaval, war, famine, or in some other way. There is little evidence at present to indicate that control of reproduction will become widespread enough in time to prevent a disaster.

DISCUSSION

POLITICS, ECONOMICS AND NUTRITION PROBLEMS

Question: Are not nutrition problems so completely involved in the economics, the traditions, and the political system of a society that they can only be solved through action by a central government? Was the pellaga problem in the southern United States solved through private enterprise? Who put vitamins in the wheat flour?

Dr. Sebrell: First, to answer your question about vitamins in wheat flour, this was the result of collaboration between industry, government, and scientists as represented by the National Research Council. To answer your question about the necessity of central government action, I would say that there something else is required that most governments do not have. It is necessary to have cooperation and coordination in national programs of health, education, economics, and agriculture if the problems of malnutrition are to be solved.

Pellagra disappeared from the southern United States as a result of improved economic status, developing business, diversified agriculture, changes in the tenant farmer situation through government support and policies, and because vitamins were added to bread, flour and cornmeal. An entire community development program is necessary to combat malnutrition successfully. It cannot be done by central government action alone.

TECHNOLOGICAL BREAKTHROUGHS IN NUTRITION

Question: Assuming that there is not very much that can be done to change rapidly the political and social institutions that militate against the adoption of adequate nutrition programs, can we spot those areas where it may be possible to make technological breakthroughs or some scientific discovery that will allow us to do something rather revolutionary within the extant legal, economic, and political context?

Dr. Sebrell: We certainly can. The application of our knowledge of food technology is one of the major factors that can make an important difference in spite of the social problems. New ways to use local foods and to preserve them so that the crop year can be extended can be very useful. Also, there is an enormous loss of food due to insects, rodents, and other pests, which could be greatly reduced by proper use of known technical methods. These are just two examples.

Question: Do you, as a medical man, say that the solution to this nutritional problem is entirely in the hands of the social scientist?

Dr. Sebrell: Not at all, although the long-range solution is certainly largely in his hands. However, the physician, health officer, and food technologist must supply him with the necessary data and tools to do the job. At the same time, illiteracy and ignorance, two very important factors must be overcome by education as part of a total community program.

Question: The situation where people are unwilling to consume the fish flour that could help prevent malnutrition interests me. Dr. Sebrell notes that "something is not working here." In other words we have discovered something which is technologically but not socially feasible. I cannot avoid comparing this situation to the Surgeon General's report on smoking. In the United States we put a certain premium on the value of health, but when we are presented with a series of actions which, logically, we should take under the circumstances, we just do not react in a completely logical way—and this in a society which prides itself on its rationality. When we consider Chilean society, perhaps we could partially explain why the population acts as it does if we look at the premium placed on health. How can the issue be dramatized for the Chilean people when the question is not directly one of life and death, but rather pertains to a substandard level of health? Do we manage to dramatize the problem so that health becomes important? How can we interest them in health to the point where they will be willing to take the necessary steps—either in the production and consumption of food or in birth control programs —to better their condition?

Dr. Sebrell: This touches upon a very important point. People usually do not consciously eat for health reasons. We eat a dish because we like it. If it is good for us, that is an extra dividend. Therefore, the problem really is: How do we get people to like a new and unconventional food? We need some Chilean food manufacturer who will take this product and sell it. The proper approach is to test the product, to test the market, to find out what will work and what will not work, and to do these things with typical business techniques.

Question: I am particularly interested in the relation between the technology of nutrition and marketing techniques. Dr. Sebrell has said that we know a good deal about the technology of nutrition, but we need to develop all sorts of techniques in social science in order to apply this nutrition technology. But it seems to me that we already have these techniques. They are found in the United States in everything from marketing Coca-Cola to conserving our forests through Smokey the Bear. It is implied that food habits are difficult to break, but twenty-five years ago we did not have pizza in this country. The question seems to me to be really this: How can we apply the techniques of social science we already know very well, together with the production techniques, to a particular situation such as that of the Chileans and fish flour? Is there any systematic way in which an inventory could be made of the taste problems that might arise in the use of protein food supplements? If so, is there any mechanism for transmitting this data to the research laboratories of the chemical technologists who work on these food supplements?

Dr. Sebrell: This is really a very complex problem. Even within the same country, different areas have so many different likes and dislikes that we have not been able to apply systematic techniques, as industry would do, to find out whether a proposed product is acceptable before it is made. If we can persuade people that this must be done, then, perhaps, we could get them to investigate, in local and specific situations, what a given population will or will not eat. The problems involved in making a new product acceptable must be studied during the development of the product. We have sufficient knowledge of technology and chemistry so that making the product meet nutritional specifications is relatively simple, compared to making it acceptable.

Question: Isn't it feasible that a case by case investigation be made, at an early stage, of the sort of taste obstacles likely to occur? And once established, couldn't these problems be specified in reports so that a commercial or international laboratory might carry out the research taking this background information into consideration?

Dr. Sebrell: It is feasible to do this, but in practice we have tended to approach the problem backwards; we worked to get a product of good protein quality first, and then tried to get people to accept it.

Question: With regard to training agents who can teach the principles of nutrition to underdeveloped countries, are there any parallels to this kind of program in the agricultural extension service in the United States, which proved to be so effective in bringing about an agricultural revolution in the twentieth century? Is it possible to utilize some of the techniques developed in this extension service to meet the nutrition problems of underdeveloped countries? Is the Food and Agriculture Organization doing anything to simplify these problems along the lines that have been suggested tonight? It seems to me that one of the things it could do is to undertake a survey of nutrition problems of the world, country by country, so that we would possess a picture of the nutrition problem of the world in concrete and specific terms. Would not this be helpful in promoting a coordinated world program?

Dr. Sebrell: There are three United Nations agencies that are most interested in these problems and are actively working on them, the Food and Agricultural Organization, the World Health Organization, and UNICEF. I am chairman of the Protein Advisory Group, which is an advisory body to all three agencies. We have not only been trying to have the three organizations carry out individual programs, but have also tried to get the three agencies together on a program for any given country.

Here are some specific projects and some of the problems that arise. UNICEF is trying to establish milk plants in areas that have never had any milk. This involves assisting governments with money and establishing dairy herds and milk drying plants, in order to make dry skim milk available. These efforts have been successful in some places, unsuccessful in others.

The Food and Agriculture Organization is working on school lunch programs, on school gardens, on fish ponds, and on other projects of this sort, but with limited resources and limited personnel. The World Health Organization is working with health departments and clinics. I should add that UNICEF is also supporting the development of some of the protein food mixtures that I mentioned in my talk.

It seems to me that the essential element lacking in all this is commercial participation. The new food products get into regular commercial food channels with considerable difficulty. They are government subsidized products distributed largely in schools or institutions. If the projects are to be really successful, these foods must eventually become commercialized, everyday foods.

Question: Could you comment on the controversy between the Food and Drug Administration, the Department of Commerce, and the Bureau of Fisheries over the use of whole fish products?

Dr. Sebrell: The Food and Drug Administration is adamant that any product made from whole fish will not be approved for sale in the United States. They assert that such products come under the classification of a "filthy" or "unaesthetic" products. Actually such products are perfectly safe, but approval of such products would break down the Food and Drug Administration's regulations regarding filthy, unacceptable, or unaesthetic products.

Of course, the Department of Commerce and the Bureau of Fisheries are very anxious to develop a fish protein concentrate and Congress has appropriated quite a bit of money to develop fish products. As I see it, it probably will require congressional action to resolve this conflict.

LAND REFORM, DISTRIBUTION PROBLEMS, AND FOOD SUPPLIES

Question: I have recently been in India. The Indian rice crop had just come in. The southern part of the country was over-flowing with rice whereas the northern part was in short supply. The distribution problem in this case was partly the result of the fact that the individual states are loathe to share their supplies of food with other states.

At the same time there has been a vast program of redistribution of land to the peasant classes in, approximately, ten-acre plots. My impression is that these small farms are doomed to be relatively unproductive. Yet here is a case where, in the name of land reform, the nation may have been penalized for a long time to come because of the problems that beset a small landowner.

Dr. Sebrell: We need a much more enlightened approach to land reform than some developing countries have shown. One thing reformers must recognize is that laws must be made to prevent the breaking up of land into very small parcels. An attempt has been made to set up cooperatives in some cases, but cooperatives will not solve the problem if the parcel of land is too small to produce enough to feed the family. At the same time a system must be developed that will enable an enterprising farmer to buy land on long-term credit and at low rates of interest, so that he will have an incentive to produce more food.

I would say that the problem of food distribution in India is even more serious than you have indicated. The Indian government seems to be unable to control food-hoarding by those who handle large amounts of rice and similar products commercially and seems also to be unable to solve the distribution problem between states.

Comment: It should be emphasized that the food crisis in India is a long-term crisis. Though people are not starving right now, the birthrate exceeds the increase of food crops.

Dr. Sebrell: That is just the point that I have been trying to make. Nobody is starving in India in the sense that he is going around with pains in the stomach, which drive him to eat grass. Nevertheless, if we compare the rate of infant mortality in India with that of the United States we find that it is roughly forty times as great, and this is largely because the Indian babies are dying of protein-calorie malnutrition. They are dying of starvation all right, but it is disguised. It should be added, however, that if the present rate of population increase continues, it is almost certain to result in some real famine in India in the next ten to twenty years.

Comment: It is not the land reform program by itself, or the socialization of land, which makes for agricultural inefficiency or a decline of food production. Rather, it depends largely on how the land reform is carried out. If an infrastructure is provided, as was done, for example, in the delta provinces of South Vietnam, there will be fantastic production on relatively small plots of land.

Dr. Sebrell: You are perfectly correct about the necessity of having some program other than that of simply breaking up large estates. When some of the large cotton plantations were broken up in the American South, an integrated program was developed in which people were educated to produce garden crops for their own needs and then to produce a money crop.

Question: In those areas where land reform has been more or less successful the condition of success may have been that the agricultural hinterland became incorporated into the money economy. Road systems were opened up; jobs were offered to the peasant population so that money began to penetrate to the rural areas. It was really the supplementary sources of income, rather than the specific land reform, that gave the population of the hinterland a chance to become more productive. I would argue that it was the WPA and similar institutions, much more than agricultural reform, which set the South on the path to expansion. It made it possible to drain the surplus rural population off to urban areas throughout the country, so that the remaining rural population was, in most places, small enough to be supported by agricultural activity.

Dr. Sebrell: Perhaps it would be worthwhile if I were to describe the agricultural problem in the American South as it struck the nutritionists during the 1920s, when cotton was still the major crop. In 1929, a vitamin deficiency disease called pellagra killed 7,000 people; there were about 200,000 cases in all. It was possible to plot the occurrence of this disease according to the price of cotton. If the price of cotton was high there was no pellagra because the farmer could then afford to buy food. When the price was very low there was no pellagra either, because the farmer did not plant cotton and instead grew food, so that his diet met minimum standards. When the price of cotton was marginal, the small and inefficient producer would plant cotton and go into debt. Credit became tighter, his food supply became more restricted, and pellagra was the result.

All this has disappeared, and it has disappeared partly because of the programs that made crop diversification possible, made it possible for the farmer to secure adequate amounts of credit, and enabled him to buy his own land. I should add that the conquest of this disease was also partly due to improvements in food supply and to general economic conditions. Industry did move into these areas, and the farmers did enter into the money economy.

Incidentally, improvements in the food supply of the South took place in remarkably simple ways. The pellagra-preventive, vitamin A, was put in all the white wheat flour and corn meal that was being sold in the South. Another important development was the spread of frozen food lockers. Previously, hogs could only be killed in the winter, and what was not immediately eaten had to be salted or smoked. With the frozen food locker, it was possible for a farmer to raise animals, kill them at any season, and freeze them. He could also sell these items, and this helped to develop a market for a wider range of high quality foods.

Comment: I have sensed that a number of people feel that the breakup of large estates interferes with technological progress. Historically speaking, the breakup of large estates has in many cases actually led to an increase in production. Large estates can be farmed in such a way as to be extremely unproductive, and, frequently, individual proprietorship even on small scale plots can provide a strong motivation to increased output. We must realize that there are two sides to this coin.

LIMITS TO THE POPULATION EXPLOSION

Question: As the picture has been presented to us, the core of the problem is the population explosion.

Dr. Sebrell: We already possess adequate techniques for controlling population; the problem is that people are not motivated to use them. Japan is the only country I know that has what approaches a successful program of birth control. This program relies very heavily on abortion. For various complex reasons the rest of the world will not adopt this method as yet. I want to emphasize, however, that if we do not voluntarily control the population growth, something else in the biological cycle is going to hold it down.

I do not think that the limiting factor will be the food supply; something else will emerge first. It may be a virus epidemic that we cannot control, or it may be a serious war, since the tensions between people are going to rise with the increasing population. I do not know what is going to control the level of population, but I feel sure that it is something that will operate before the food supply runs out, because we do have the techniques for enormously increasing the supply of food, even if we do not at present have adequate techniques for distributing it. I do not mean to say that famine will not occur. Some famine is almost certain to develop in the lower economic groups of the developing countries, but this is not likely to be severe enough or widespread enough to precipitate a solution to the population problem.

Question: Dr. Sebrell's position is that we can solve the food problem for the foreseeable future until some other factor limits population. What Dr. Sebrell has said is, fundamentally, that among the three positive checks noted by Malthus, starvation is not likely to be the first to operate. We are likely to have either disease or war first, but this does not alter Malthus' analysis of the situation.

Some Speculations on the Social Impact of Technology

by DONALD N. MICHAEL

*Resident Fellow, Institute for Policy Studies,
Washington, D. C.*

I WANT TO DRAW your attention to what I think are the important aspects of the social impact of technology that we have so far ignored or attended to only in passing. I do not intend to convince anyone of anything: I am not sure what the questions are in this area—much less what the answers might be. Rather, I want to sketch a variety of perspectives and circumstances that I think merit attention, so that you may determine whether or not these are significant issues for scholars and actionists concerned with the impact of technology on society on more than an occasional basis. These are issues that we ought to be concerned with at least as much as we have been with productivity, investment policies, employment, and so on.

Let me begin by making clear that I am not asserting that technology is a villain, or that technology is a saviour. The problem is not that simple. The positive and negative interplay between technology and social processes is much too complicated to comprehend if only one aspect, technology, rather than technology in the context of society as a whole is dealt with. In no sense am I insensitive to or unappreciative of the great opportunities implicit in technology, particularly in the new technologies. Nor may the odds on adverse consequences from these technologies be any higher than on favorable ones. But I believe that the consequences themselves, favorable or unfavorable, are of such magnitude that if they are negative, they will bring upon us much more serious trouble than we would have had in the past, in simpler days, when technologies had fewer derivative implications, and effected fewer people in a smaller area over a longer period of time.

Let me also make clear that I do not believe that the solutions to the problems we will discuss are to be found, by and large, in a moratorium on technological development. Such is the social environment technology has already produced that, unless we are to change our value system and way of life totally, we must use more technology to make an adequate environment out of the circumstances technology has already brought about. I fully expect the technologies to help deal with the problems the technologies produce. I also fully expect that unless we take a larger and deeper view of the social implications of technology than we have so far, we will not use our technologies or other resources sufficiently to protect us from the enormous potential for social disruption and disaster implicit in these technologies. Hence, I am going to emphasize aspects of the social impact of technology that I believe present problems which must be solved if we are to enjoy the advantages the technologies can provide. I think we can expect in our type of society, and in a society as rich as ours, that the opportunities will take care of themselves. Put another way, the opportunities do not need to be optimized, but the dangers, I think, must not be minimized.

Sketchy and impressionistic as these observations will be, I have tried to organize them into three general categories. First, I shall make some observations on the general considerations that should be applied to any estimate of the present and contemplated social impact of technology. Then, I want to mention three technologies which I expect will have a wide social impact in the next two decades or so, and which by their characteristics imply a far broader range of social impact than we have felt so far, certainly than we have studied so far. And finally, I will set out some examples of aspects of social impact that should be studied intensively now, if we are to be prepared to use the results of such studies for guiding the felicitous integration of technology and society in the years ahead.

Let me say something else by way of introduction. In many ways I will be implying that the methods and knowledge of our various disciplines are inadequate or nonexistent and, therefore, cannot help us understand what is happening to society vis-à-vis technology. I hope I will be able to imply convincingly that many, probably most, of what may be the significant issues are *not* being explored effectively and on a scale and with the attention they deserve. If I "get to you" I will thereby, inevitably, threaten our various senses of self and status and purpose; I will question who we are, what we do, why we are what we are, and how important we are to ourselves and to others.

There are typical ways to defend oneself against such threats: by "not hearing" or misunderstanding the speaker's choice of points of emphasis and context of qualifying remarks; by translating and transforming what he emphasizes into a problem or a syntax with which the auditor is comfortable and familiar, thereby shifting the plane of discourse; by attending to the speaker's mood rather than to what he says, and so on. Inevitably, these defenses will operate here just as they do throughout the community of persons and institutions whose favored perspectives, and, thereby, senses of self, are challenged by the interplay of technology and society which makes obsolete or inadequate the conventional techniques for perceiving the society and dealing with it. Indeed, this type of threat to self and these responses to it, produced by a changed and changing world, conveyed from one person to another through the different perspectives of those involved, are in themselves important social impacts of technology about which we know too little, and which we need to understand much better.

With this forewarning, let me now turn to some general considerations on understanding the relationship between technology and society, which I think are too seldom appreciated or made central to the context when specific issues are explored.

1. It is important to remember that some of the significant impacts of technology derive from the accumulated effects of technological changes that have been under way for some time. Let me remind you of three examples: the population explosion, a direct product of medical technology initiated some years ago; the urban chaos that has been fundamentally exacerbated by transportation technology in the form of the private car; and the distortion and disruption of ecology in local areas and, probably, in much larger areas by the wholesale application of pesticides and fungicides, to say nothing of the continental pollution of the ecology by the waste products of many technologies. In the future some of these accumulated consequences will become more emphatic and complex, when the new technologies contribute their consequences too. The technologies we have to be concerned with thus include some older ones as well as the new ones. This is important: we do not have to wait for the new technologies in order to improve our concepts and methods for understanding their impact. All we have to do is recognize our ignorance and indifference regarding the present social impact—and do something about it.

2. A major purpose of our preoccupation with technology and social change is to prepare for the future. But doing so is going to be very difficult. There are some social consequences of technology for which we should have begun to prepare yesterday. For example, the type of education appropriate for a rapidly changing work force and for substantial increases in leisure probably requires fundamentally different attitudes and approaches by the teachers in primary and secondary schools than

the present ones. But recruiting the teachers and teaching the teachers requires changes at least in the schools of education and in teachers colleges, and all such sequential changes take time to introduce. The upshot is that most of those teaching today's youngsters and imbuing them with the values and attitudes they will carry into their lives tomorrow are probably conveying the wrong things, because the teachers were wrongly trained and perhaps, in part, wrongly selected. Similarly, problem-solving investigations in the area of urban affairs must be undertaken long in advance of the actual reconstruction and reorganization of our cities. The riots in Los Angeles and the water shortage in New York may be mild precursors of the potential disasters that may otherwise overtake us.

In general, we do not understand and appreciate, and thereby tend to overlook, the nature of the time lag between recognition of a problem and the development of techniques for dealing with it.

We do not allow for—often we do not know how to allow for—the needed intervening period to accumulate knowledge and understanding. I suspect that over the next couple of decades or so, this gap between problem recognition and the development of solutions, or approximate solutions, is going to become increasingly serious: problems will confront us before we are able to deal with them knowledgeably. The integration of technology with other social processes and the felicitous sequencing of contingent social actions to accomplish the integration is going to be much more difficult than it has been in the past.

3. Value conflicts, and tensions between generations will very likely increase, especially between the new generation that is moving into political and professional power and is using new types of operational and substantive expertise, and the older generations already occupying the field. These conflicts and their various expressions in differing values and operating

techniques will mean that both the pressures and the inhibitions to make the kind of social and technological changes that we are going to need are certain to be very great. As the distinguished public servant, systems theorist, and student of decision processes, Sir Geoffrey Vickers, puts it:

In the transitional period from the conditions of free fall to those of regulation (at whatever level), political and social life is bound, I think, to become much more collectivist or much more anarchic or —almost certainly—both. Communities national, subnational and even supranational will become more closely knit in so far as they can handle the political, social and psychological problems involved and more violent in their mutual rejections in so far as they cannot. The loyalties we accept will impose wider obligations and more comprehensive acceptance. The loyalties we reject will separate us by wider gulfs from those who accept them and will involve us in fiercer and more unqualified struggles.[1]

What the resolution or stalemate will be between the old and new approaches to the uses of knowledge and power via technology remains to be seen, but an understanding of this conflict in approaches will be prerequisite to an understanding of the social impact of technology.

4. Under what circumstances does an issue become significant or critical? What determines when small percents, such as the unemployment rate for example, become sufficiently large in absolute numbers to become a major issue? When does indifference become transformed into action, as for example, in the poverty program? When does the ability to extrapolate trends begin to carry real significance in terms of program implementation, as, for example, in the moon program, where the ability to use computer technology to predict whether we could succeed was an important factor in the decision to go ahead.

Understanding the general principles of when, or how, or

[1] "The End of Free Fall," p. 21. Mimeographed article (Fall, 1964).

why issues become important and become recognized as issues, would be essential for estimating the seriousness of or coping with, among other things, the gap between posing a problem about and finding a solution to an impact of technology on society.

5. Often interpretations of the past are called upon to help interpret the present and to suggest solutions to expected problems in the future. By and large I think these interpretations have been inadequate. First of all, there is often only a surface similarity: the presumed analogy is based on a partial or on a misunderstood picture of the past. A prime example of this is the frequent submission of the Periclean age of Greece as evidence of what our future leisure patterns should or could be. That the "leisured" society of Greece numbered only in the tens of thousands, that it spent much of its time warring or doing those political tasks we have professionalized, that no women were involved, that the system lasted only a couple of generations and then decayed, all such considerations are left out of the supposed apt and happy analogy.

Or consider the argument that since we mastered the first Industrial Revolution, we will not have any enduring trouble with the second industrial revolution that the new technologies represent. But we have not mastered the first Industrial Revolution. Let me mention three social consequences of the first Industrial Revolution that are increasingly acute. While other factors have contributed to them and while in some form or another they may have existed previous to the first Industrial Revolution, undoubtedly our inability to deal with the consequences of that revolution have enormously exacerbated these conditions. The first is the increasing gap between the developed nations and the have-not nations that was widened by the technological prowess of the developed nations, and by their inability to share their technology with the underdeveloped nations. The second example is the persistence—the growth, in some cases—of slums and poverty. This peculiar type of degrading, enclaved existence was the direct result of the first Industrial Revolution's concentration of factory technology, and the resulting transfer of manpower from the farms to the factories. The third example is the alienation from and breakdown of earlier, more stable, systems of value and faith. There are very few students of this problem who feel that the Industrial Revolution has not contributed enormously to the complexity and persistence of the problem.

Another inadequacy in appealing to the past is that even if there is more than a surface similarity in the particular social processes involved, there usually are differences in the surrounding social or physical circumstances, which imply very different consequences than those occurring in the past. Two very important examples of such differences are worth mentioning.

First, never before in history has any nation had such a complex technology combined with a population as large as that of the U.S., now and as it will be—230 million around 1975, 250 million around 1980. Those who assert that because thus and such a technological consequence was coped with in the past, or that a specific social consequence is not new and neither, by implication, are its future consequences, have to consider whether the multiple social and physical consequences of a population of such huge size carry significantly different implications for the future.

The second difference, which we tend to overlook, is that today there very probably are different expectations than there were in the past of what the implications of technology on society will be. Because people are different today, they expect different things to happen to them as a result of technology than people did in the past. It does not help to say, "Their beliefs do not really jibe with the facts." (Especially since we probably do not know the facts.) If they think, for example, that technological change is galloping along at a rate never before equalled, then this results in various reactions by business executives, scientists, pundits, and government officials— to say nothing of the man in the street. These reactions, very likely, are significantly different from those in the past, when this kind of issue did not cross most people's minds. For example, I strongly suspect (but we have not bothered to find out) that a recent major social consequence of the interaction of several technologies is that more types of people are concerned about the future social impact of technology, and more different viewpoints have arisen regarding it, and that these in turn are initiating other social consequences.

Let me now turn to three technologies, biological technology, cybernation, and social engineering. I expect these will, over the next two decades, have very great implications for the nature of society and for the place of the individual in it. I want to emphasize those implications that go beyond those we have tended to preoccupy ourselves with, not because I think they have been unimportant, but because I believe the ones I want to mention are equally important, perhaps more so. Let me again make clear that I recognize that many of the impacts will not be new in kind. However, I do believe that it is likely that the scale and scope, the potency, of their impact, as they interact with an already huge and enormously complex society, will be of an unprecedented order of magnitude. In that potency of impact lie many exciting opportunities, and some very profound problems having to do with the place the individual has in a democractic society, and the way we conduct our lives. Again, I will emphasize problems not because I am certain that they will outrun the opportunities, but rather because I feel that if the problems are not dealt with effectively, the consequences may be so disastrous that we shall never enjoy the opportunities—at least not within the format of presently preferred values. (Of course, a different set of values may be all right or even better than the present ones. But what I want to emphasize is the kind of confrontations these technologies are to our present values, if only to indicate the necessity for understanding the social impact of technology better, so that we may invent or respond to a more appropriate set of values.)

Perhaps the fundamental question that the potency of these technologies raises is how do we deliberately decide how we are going to balance the social costs and social benefits; obviously I mean infinitely more here than the dollar costs and benefits. I think we must do much more than hope for the best or retreat behind some inhuman "averaging out" philosophy. But what do we do if, as I believe they will, the consequences of these technologies will be upon us before we accumulate the understanding needed to establish such a balance—if we ever can accumulate such understanding?

In the application of biological technology—the engineering of man's biological self and his biological environment—we will face moral, ethical, psychological, and political issues, which will make those faced by the atomic scientists look like child's play. Biological and chemical warfare will very likely be used much more in local wars, even perhaps in the pacification activities of international police forces. But whether it is used to kill, hurt, nauseate, paralyze, cause hallucination, or to terrify military personnel and civilians, the systematic use of biological and chemical warfare will require the resolution of major moral and ethical problems—especially since the most likely victims will be nonwhites in Asia and Africa.

Psychopharmacology is another aspect of biological technology already beginning to confront us with interesting issues. What is to be the role of hallucinogenic chemicals in society? There are two schools of thought on this—even the theologians seem to have taken sides. One is that these chemicals represent sin and corruption; the other, that they are exciting means for enlarging emotional or aesthetic or religious experience. Moreover, new drugs will permit many people, who otherwise would be in mental institutions, to walk the streets and to engage in regular social activities. Questions arise about the "nature" of the individual. How do we judge the extent to which a person is "responsible" for himself in such circumstances? For, while the chemical effects the individual, the person is significant to himself and to society in his *social* context—at work, at home, at play. The consequences are social consequences. In deciding how to deal with such alterers of the ego and of experience (and consequently alterers of the personality after the experience), and in deciding how to deal with the "changed" human beings, we will have to face new questions such as, "Who am I?" "When am I who?" "Who are *they* in relation to me?"

As far as the hallucinogenic agents are concerned, how will we judge whether people, just because they want it, are entitled to a risky if richer emotional experience than that provided by their everyday life? Are these decisions to be left to the individual, like skiing or surfboarding, or will they need legal restrictions like homosexual liaisons or the present use of nonhabitforming marijuana? In general, will "multiple" personalities and increasing amounts of idiosyncratic behavior simply be absorbed into the already proliferating scale of novelties, sensations, and leisure-time pursuits, or will they have to be controlled to facilitate the functioning of a stable society? Whatever way is chosen, what are the ethical, legal, political, and psychological considerations needed to help us understand the implications of such altered egos and their control?

A related aspect of biological technology merits mention here: with the increasing dissemination of birth control information and technology, we can expect the pressures on the poor to limit family size to become greater. Though such pressure already exists, in the form of inadequate housing for large, poor families, the pressure may well become more explicit as the "excuses" for having large families inadvertently are eliminated by the pervasive availability of birth control methods. If our laws and ways of operating come to condone this invasion of the right of couples to choose the number of children they want, then a new ethical issue will arise and it will reverberate into other areas of private affairs, conduct, and choice.

A third area in biological technology has to do with organ transplants. Some top research people in this area are convinced that in a few years techniques will have improved substantially. The problem then arises, Who is entitled to what transplant under what criteria of priority? We will have to do better than "women and children first." This situation already exists on a tiny scale with regard to the use of scarce kidney-substitute machines. Difficult as the decisions are now, they will become more difficult and more socially consequential when more people compete for more organs.

Though it is unlikely that organ transplantation will be available to such an extent as to increase substantially the number of people who will live longer, it is likely that developments in the technology of preventing and treating malignant diseases, will mean that there will be still more older people in this society, which has not yet learned to deal humanely with the older population it now has. Thus, developments in biological technology, combined with those from cybernation, in particular, will add to the numbers and to the social problems of older people. The accumulated social impact on our political system, and thereby on our social priorities, will undoubtedly be substantial as the old become a larger proportion of the voting public. Here, too, our understanding of the situation is much too slight at present to give us the knowledge we need to plan effectively for this growing population.

Finally, there is the question of genetic engineering: the deliberate controlled alteration of human inheritance. Late in the next couple of decades, either the capability to do so will exist or almost certainly it will become clear that soon thereafter the capability will exist. Indeed, there are already expressions of exuberant optimism, as well as sober concern, about the possibilities this presents. The optimists, typically, are concerned with technological manipulation, pointing out that maybe we could give everybody an IQ of 140 and eliminate all inherited human "inadequacies." The concerned, typically, look at the looming social and ethical issues that arise from such actions. For example, what are the psychological and, therefor, social consequences of producing a generation of adults who, as youngsters, shared little with their parents because their IQs were so much higher? And who is to decide when an inherited "inadequacy" is one that should be eliminated by genetic engineering? Who will decide where the line is to be drawn on the definition of "inadequacy"? Fundamentally, who will make what decisions about which human beings are to be changed before they are born, and in what way? Or, for that matter, who will determine that we will not use the technology with its implicit potentialities for improving the race?

Cybernation—the application of automation to material processes and the application of computers to symbols—is the second technology I want to mention. I shall not dwell on the usual questions about cybernation's impact on employment: they have been discussed amply enough to demonstrate our awareness of the matter as a social impact—even if we are unclear on what the impact is, much less what it will be. (Indeed, I suspect we put so much of our emphasis on the employment effects of cybernation simply because, having some figures and concepts available, it is psychologically more comfortable to emphasize this narrow aspect of the issue than to struggle with the clear evidence of our wider ignorance.) However, two aspects of cybernation's effects on employment should be mentioned here to broaden the picture.

Substantial numbers of the relatively skilled, including the middle-level manager and the middle-level engineer, are going to be displaced; *Business Week, Newsweek,* and *Time*—a little late—acknowledge this.[2] The competences that have made these people economically valuable in the past will increasingly be made obsolete, either because cybernation, particularly computers, can do the job better, or because the process of rationalizing the overall activity in which they were involved will eliminate, or substantially reduce, the need for humans to do the tasks. Here we have members of a career-oriented, affluent segment of society, who were brought up to believe they possessed all the credentials for a lifetime of advancement, now forced to find another job, or to go back to school and learn something new. They are now perpetually under the threat of being displaced by younger men and by more sophisticated machines. Many of these people are already anxious and insecure personalities—as well as substantially in debt. It is likely, then, that they and their families will suffer considerable disruption as they revise their images of themselves; who they are, what they might become, and how others see them. What will happen to these ex-cynosures, to their aspirations, and to their way of living? And what political action will they take in response to the threats to their status and security?

There is another economic and emotional problem that cybernation's impact on employment level and employment changes will pose: What is to be the future of unskilled women in the work force? In the work force, of which one third are women, about 80 percent are no more than semiskilled; many of them do the clerical and routine service jobs, which cybernation will replace as its application is accelerated by the increasing size of organizations responding to the increasing population. Now, about 60 percent of the $9,000 to $15,000 a year incomes in this country are those of families in which both spouses work. If

---

[2] "Computers: How they're Remaking Companies," *Business Week* (Feb. 29, 1964); "The Challenge of Automation," *Newsweek* (Jan. 25, 1965); "The Cybernated Generation," *Time* (April 2, 1965).

the unskilled women lose their jobs, there will be less family income, less consumption. And there is also the question of psychic income; many women work for other reasons than to earn money. What will provide this psychic income?

Doubtless, many jobs could be invented, particularly in the human services area, which trained women could fill and which, because of the interpersonal nature of the task, no machine could do. As it now stands, however, and the poverty program demonstrates this, we are neither seriously inventing these jobs nor making the elaborate effort needed to motivate young women so that we can retrain them, and older women, for such jobs. It is likely that this problem will be upon us before techniques for job-producing, motivating, recruiting, and training are sufficiently developed, in which case there will be serious social consequences. But if we do develop such techniques, society will become significantly different from today's, because the roles of so many women will be so different from what they traditionally have been.

This leads me to the third technology: social engineering. I yield to no one in my reservations about the ability of the behavioral sciences to deal with complex issues at the present time, but the evidence to date indicates that this situation will very likely change dramatically in the next two decades. The combination of large research funds and the computer provides the social scientist with both the incentive and the technique to do two things he has always needed to do and never been able to do in order to develop a deep understanding of and technology for the manipulation of social processes.

On the one hand, the computer provides the means for combining in complex models as many variables as the social scientist wants in order to simulate the behavior of men and institutions. In the past, the behavioral scientist simply could not deal with all the many important variables that would help him understand and predict human behavior. Now he can.

(This is not to say that everything that is important about the human condition can be so formulated, but much that is important can be put in these terms; almost certainly enough to bring about substantial improvements in our ability to understand and predict behavior.) And then the social scientist can test these models against conditions representing "real life." For, on the other hand, the computer has a unique capacity for collecting and processing enormous amounts of data about the state of individuals and of society today—not that of ten years ago. Thus, the behavioral scientist not only can know the state of society *now*, as represented by these data, but he can use them to test and refine his theoretical models. The convergence of government programs and the computer is of critical importance; it will result in an efflorescence of longitudinal studies of individual and institutional change as functions of the changes in the social and physical environment. Such knowledge, now essentially nonexistent, will inevitably increase our ability to effect social change. And given the convergence of the powerful technologies and our already enormously complex and huge society, it would seem that social manipulation will be necessary if we are to introduce appropriate changes in society at the appropriate times. The problem, of course, is: Who is to decide who is to be manipulated and for what ends.

Let me now turn to some general questions regarding the social impact of technology—questions that to some extent refer to circumstances already with us, and which seem to me to be greatly in need of serious and extensive study. Let me hasten to add that I am certain that in many cases we do not at present know how to study these problems, but if we do not start now to try to invent means for doing so, we shall be in a far worse position when the time comes for us to understand these issues better. Again, the time lag problem bedevils us.

What happens to the sense of self in a world of giant and pervasive man-made events, especially when, at the same time, we insist on emphasizing the autonomy of the individual? We talk about the importance of the individual and of the wealth of options this world offers him. Yet, we have surrounded him with pollution, radiation, megalopolis, etc., which, though man-made, may appear to many people to be of such power and scale as to dominate them like "acts of God." How does a man see himself in relation to his espoused ideal of individual autonomy when he also sees other *men* and man-made circumstances, as awesome and implacable and often as impersonal as "acts of God," framing his destiny?

What kind of personalities live most fully in the midst of multiple and simultaneous change? Daniel Bell has pointed out that we are experiencing the end of the rational vision, that events today (and more so tomorrow) do not have simple cause and effect sequences, that, instead, events all happen at once and in circular and probabilistic ways.[3] What kind of person can live meaningfully in that type of world, and can keep in touch with it?

I suspect three kinds of responses will have increasing social implications as technology alters the scale of events that define the individual to himself—and thereby the ways in which he responds to the world.

One response is that of "selective involvement." People pick the issues and things they are going to respond to and be responsible about, and ignore the rest. We know people do this now, deliberately or, more often, unconsciously: there are limits to the amount of information humans can process in a given amount of time.

Therefore, it behooves us to examine carefully the degree of validity, as measured by actual behavior, of the statement that a benefit of technology will be to increase the number of options and alternatives the individual can choose from. In principle,

[3] Daniel Bell, "The Post-Industrial Society," in *Technology and Social Change*, ed. by Eli Ginzberg (New York, Columbia University Press, 1964), pp. 58-59.

it could; in fact, the individual may use any number of psycho-logical devices to avoid the discomfort of information overload, and thereby keep the range of alternatives to which he responds much narrower than that which technology in principle makes available to him.

Another type of response, now evident among returned Peace Corps volunteers, college students, and some executives, is withdrawal—pulling out of the big system, looking for environ-ments in which one can have face-to-face relationships in a simple, less technologized, more direct world.

A third response, protest, is exemplified by such things as the urban race riots and the Berkeley demonstrations. Here, the individual responds to overwhelming complexity by side-stepping the legal or ethical constraints that sustain or are at least associated with the complexity. (It is worth noting that a battle cry in the Berkeley protests was "Put your body where your punch card is!" It was one of the chief reasons for the sit-in in Spraoul Hall.) I suspect that these attempts, these experiments, to simplify an increasingly complex world will have very important social consequences, produced, in part, by a proliferating technology. If these responses are important in the future, we need to know much more about them, at least as responses to technology, than we do now.

Another way to look at the implications of technology for the individual is to consider the roles he plays. Two examples typify the unanswered and, for the most part, unstudied, ques-tions in this area. The psychiatrist Robert Rieff has suggested that to the extent that tomorrow's society is service-oriented (material productivity becoming increasingly cybernated), many men will play roles which traditionally, in our society, have been women's roles, i.e., person-to-person helping roles. What, then, happens to the image and conduct of men? What happens to the relation between the sexes, as the hard won pattern of women competing with men for "male" jobs is reversed, and men begin to compete with women for "female" jobs?

A second role implication: as society puts more and more emphasis on rationalization, logic, science, and technology, and as our educational system reflects this emphasis from the lower grades on, what will be the role of the mother—the female— in preserving the ineffable, the intuitive, and the aesthetic in the basic learning experiences of the young? This, traditionally, has been what we expect of women, but traditionally we have deprecated those contributions, at least out of one side of our mouth. Will we come to appreciate this contribution more? Will we insist that women fulfill this role more effectively, or will we further deprecate its utility in a society oriented toward technology? And what effect will our choice have on our way of life and on societal goals?

The opportunities and problems that increased leisure—resulting from the increased productivity of the new technologies —provides, to help individuals find themselves or to extend the means by which they lose themselves, have been commented on extensively and, to my mind, unimaginatively and unperceptively. I will not explore the issue here. A couple of observations, however, are in order.

An increasing number of theologians and religious denominations are becoming concerned with this problem. Their theologies assert that it is through work that man gains his salvation and fulfills himself. If work is to be a much less significant part of life for more people, what are the revisions in theology and the revisions in religious bureaucracies required to cope with this? On the other hand, the Protestant ethic, in its original form, may not be as pervasive as we have surmised, or at least its modes of implementation may be changing. Instead of leisure being a reward for hard work, we "travel first and pay later"—which may mean, of course, that work is now a "punishment" for taking the leisure first. Or, of course, it may mean that many people no longer need the justification of work in order to comfortably enjoy a vacation.

We can assume that leisure should have meaning in addition to that associated with recreation and hobbies, as is now taught. But it is hard to see how the state of mind required for this is to be conveyed to young people in an educational system stressing efficiency, and by adults who themselves are products of the Protestant ethic. Tranquility, contemplation, loafing, the cultivation of self, require a different school and different teachers. Just how real or serious would be the variety of social consequences implicit in these observations remains to be seen. Again we have not studied and again we have not tried to lay out the implications in a sufficiently elaborate social and technological context.

What *is* to be the relationship between the churches and an increasingly rationalized and technologized society? In a society preoccupied with dealing with the average, with the mass of the population, with grandiose schemes for remaking man and his environment (often accompanied by an arrogance indistinguishable from *hubris*), will it be the role of the churches to insist on another set of values for judging the direction and purpose of man, in order to protect the ideal of the individual and the validity of extralogical and transcendent motives and experiences? Here, indeed, a profound confrontation between two cultures *may* occur or, perhaps, one may absorb the other. Whatever the case may be, the consequences of the new technologies for the churches are bound to be great.

Consider the changing role of the scientist and the scientist-engineer. The symbiosis between science and technology has, as we all know, evolved into big science and big technology, and these two, in turn, are dependent on big money, which inevitably means big politics. The result, as a Report of the Committee on Science in the Promotion of Human Welfare of the American Association for the Advancement of Science argues, is that the integrity of science has been eroded and that in the absence of procedures (which have not been invented, much less implemented) the erosion of the integrity of science will very likely increase in the future.[4] In part, this is because in the future still bigger technological investments in science and engineering will be needed. Hence, still more funds will have to be raised, and political methods will have to be used still more often to mediate between the needs of technology, the other needs of society, and the needs of competing groups within the science and engineering communities. Inevitably, there will be persistent, very likely increasing, confusion between the political and rhetorical validity and utility of scientific knowledge and its inherent scientific validity and utility. For not only will scientists and engineers turn to politics to get the technology they want in the first place, but they will use politics to praise, apologize for, or criticize the social consequences of that technology when they happen.

[The] combination of esoteric knowledge and political power alters the function and character of the scientific elites. They no longer merely advise on the basis of expert knowledge, but they are also the champions of policies promoted with unrivaled authority and frequently determined by virtue of it. In the eyes both of the political authorities and the public at large, the scientific elites appear as the guardians of the *arcana imperii*, the secret remedies for public ills.

As the nature and importance of scientific knowledge transform the nature and functions of the scientific elites, the availability of democratic control becomes extinguished. Scientific knowledge is by its very nature esoteric knowledge; since it is inaccessible to the public at large, it is bound to be secret. The public finds itself in the same position vis-à-vis scientific advice as do the political authorities: unable to retrace the arguments underlying the scientific advice, it must take that advice on faith.[5]

[4] *The Integrity of Science* (Washington, D.C., American Association for the Advancement of Science, 1964).

[5] Hans J. Morgenthau, "Modern Science and Political Power," *Columbia Law Review*, CLIV (1964), 1402.

The growing potency of social engineering will become a crucial ethical issue for the behavioral scientist. Whether he is working for the government, for business, for Madison Avenue, for the CIA, for the Poverty Program, or is doing basic research, the results of his work are going to be used to "guide," "stimulate," "motivate," and "manipulate" society. Again, it is the potency of the technology, its capacity to do wonderful good or monstrous evil, that will make the situation in the future different from the past.

This ethical problem: whether to assist in the growth of social engineering, is going to become ever more serious as the potency of social engineering increases. And right now we have no ethical or scientific models for dealing with this problem. One example of this dilemma: The Job Corps trainees will have very elaborate computerized reports prepared about them, to cover their whole social and psychological background, their experience in the Job Corps, and what happens to them for several years after they have left the Job Corps. The reason for these records is a very good one—they will improve the selection and training techniques. However, such a record also means that the Job Corps trainee will no longer have a private life: once recorded, his life history will always be available in this form. The dilemma that faces the social scientist is that on the one hand he needs this kind of information to improve the Job Corps, and that, on the other hand, so much personal information made available to as yet unspecified people, may completely undermine the conventional privilege and social advantages of privacy.

Underlying these issues is the profoundly important one: what are the implications, for the form and conduct of democratic political processes, of the complex social issues that technology generates and of the esoteric methods technology provides, for dealing with these complex issues? The increasing complexity of social problems and of the techniques for dealing with them will mean that the average well-educated person—

to say nothing of the man in the street—will no longer be able to understand what the issues and the alternatives are.

This will be partly a matter of the complexity of the issues and of the technologies for defining, interpreting, planning for, and then dealing with, them. It will be partly a matter of the partial availability of knowledge. Often the issues will be politically sensitive and, as now, the interested parties will release only what they wish to release. Moreover, laymen able to use the knowledge, if they did have it, would need reasoning abilities which most people now lack. They would have to understand that the world picture is in most critical cases a statistical one, not black or white. These laymen would have to be comfortable dealing with multivariable problems operating in multiple feedback processes, where cause and effect are inextricably intermixed, and where it is often meaningless to try to differentiate one from the other. And they would have to be comfortable with making judgment based on a much longer time perspective than most people are used to now. They would have to be able to think ahead ten and twenty years, and make their judgments accordingly. These are not characteristics we are going to find in large numbers in our population: our educational system simply does not mass-produce such people —and evidently will not do so for some years to come. But if we are to operate a democracy, the need for such reasoning abilities will be upon us sooner than that. Indeed, it is already upon us. The political scientist and pundit, Joseph Kraft, recently observed:

To apply common sense to what is visible on the surface is to be almost always wrong; it produces about as good an idea of how the world goes round as that afforded by the Ptolemaic system. A true grasp of even the slightest transaction requires special knowledge and the ability to use abstractions which, like the Copernican system, are at odds with common-sense impressions. Without this kind of knowledge, it is difficult to know what to think about even such prominent matters as the United Nations financing problem, or the bombing of North Vietnam, or the farm program, or the federal Budget—which is one reason that most people don't know what they think about these questions. The simple fact is that the stuff of public life eludes the grasp of the ordinary man. Events have become professionalized.[6]

Moreover, the problems, whether they be urban renewal, air pollution, education for the new age, Medicare, international development programs, the exploitation of the oceans, assigning technology development priorities, etc., will be too complicated to be dealt with effectively by the techniques that have characterized our society to date. And the issues will be too critical, the potentials for and scale of disaster too great to stake our social survival upon conventional approaches—even when they are undertaken (as they rarely are) with the best of disinterested goodwill. All we have to do is to look at the looming disaster our cities represent, to recognize that we are going to have to do much better.

The tasks we face, then, will require the full use of whatever rationalized techniques we have, and these techniques will proliferate in the years ahead with advances in the social sciences, with increasing use of computers, data banks, simulation, system analyses, operations research, and so on.

In consequence, planners and decision makers will be confronted with a set of circumstances that will also suggest important changes in the democratic process. The competing demands for human and physical resources, necessarily expended over long periods of time, will require the development of ways to assign priorities and to revise costly efforts, even if it is politically uncomfortable and institutionally disruptive. At present we have neither the priority scheme nor the means for efficiently and reliably transcending conventional and institu-

[6] "The Politics of the Washington Press Corps," *Harper's Magazine* (June, 1965), pp. 101-2.

tional restraints. Yet, obviously, we will have to be able to choose between major technological and social developments, and we will have to be able to maintain or alter these decisions more in the light of their real accomplishments, rather than in the light of political commitments. Furthermore, because of the massive needs of the society, there will be a tendency to respond to average human and social requirements, rather than to the needs of the individual qua individual. This tendency will be exacerbated by the inherent characteristics of technologies, of systems analysis, and of operations research and computer simulation. The pressures to value those things most about the society that can be described and dealt with in terms of the techniques available, and the pressures to deal with the massive needs of the society will make it especially difficult for the policy maker and decision maker to preserve a sensitivity for and responsibility toward the idea of the idiosyncratic and extrarational needs of the individual.

If, then, we are to preserve the ideal of the cherished individual we will need wise men more than we will need technically skilled men, though obviously we will need the technically skilled men too. As it is, we do not know how to produce wise men, and we do not know how to provide them with an environment that will encourage their wisdom to blossom and act. Yet without wise men, the chances are that the democratic concept and the Judeo-Christian tradition built around the obligations and rights of the individual will be lost under the crush of the vast needs of the society and the enormous potency of the technologies put into operation in a massive society to meet those needs. How shall we prepare for and invent the new forms of democracy and the new roles to be played by citizen and leader in such a system?

Above, I implied the need for the ability to change institutions rapidly. This, too, is a consequence of the impact of technologies on society, for through their effects technologies make the mandates of institutions, and the validity of the operating styles within the mandates obsolete. Yet institutions persist and change only slowly and usually reluctantly—barring some kind of disaster. Some observers have pointed out the potentialities for society if we apply our technologies. They then bemoan the apathy of the public and the ineffectualness of institutions because they do not take advantage of the technologies. The usual interpretation of this state of affairs is that we lack "leadership." But this is a naïve solution and a premature definition of the problem. The question really is how to change institutions so that leadership arises in a given situation and then acts. Here our formal knowledge, limited as it is, makes it clear that this is an extremely difficult condition to deal with expeditiously.

As institutions produce and use the new technologies, they inevitably will have to change at a rate concomitant with the changes produced in the society by the very technologies they have encouraged and applied. But getting institutions or, rather, the people in them, to shift their perspectives radically as technology radically alters reality; getting the members of institutions to risk statuses, self-images, empires, in order to prepare for future needs, is an enormously difficult task, usually only successfully accomplished after a major institutional disaster has occurred. Over the years we can expect that the social sciences will provide us with more knowledge about how to make these changes quickly (or perhaps provide us with an understanding of why, if we want to preserve a humanitarian set of values, institutions cannot be changed quickly). But even if we assume the former, there are still many years ahead in which institutions will lag behind in their ability to respond to the real environment as it is altered by technologies, and this lag will become increasingly dangerous. What we do now and in the long run about this impact of technology is a matter that I believe deserves intensive attention.

Perhaps we might do well to spend some significant portion of our professional time stockpiling solutions to social problems, which we cannot hope to get into our social system now, but which we can reasonably expect to apply if some of these problems back up on us to the point where we cannot cope with them within the present social format. It is after disasters that institutions can most easily be changed.

Let me close with some comments on the special social impact on scholars or action-oriented professionals of the very *question* of the social impact of technology. One direct effect of the new technologies is to challenge deeply the adequacy of our academic disciplines for dealing with the kind of world they are producing. We sit here and talk learnedly about economic and social processes—rates of change, institutional process, etc.,—but my impression is that few of our disciplines or techniques are now really adequate. Even in the well-studied area of productivity and technological change we cannot be sanguine about our methods. As Solomon Fabricant recently said, "The problem of measurement has not yet been solved. . . . There are competing and widely differing measurements of technological change. . . . I'm afraid that people talk about both the past and the future . . . with more confidence than is warranted by the available knowledge about technological change."[7] In the few cases where our techniques are adequate, they are not being used broadly or intensively enough to deal with the multiple issues that must be understood if we are going to secure the advantages these new technologies possess.

If my impression is accurate, we face some very uncomfortable questions, which, as scholars and professionals, we are morally bound to wrestle with far more than we have until now. What about our research techniques? What must we do—and what must we abandon of what we now take status and comfort

[7] *Measurement of Technological Change* (Washington, D.C., Manpower Administration, U.S. Department of Labor, 1965), p. 3.

in—to get methods that adequately tackle the issues? We must find out what we should really be studying, even if it means breaking down cherished disciplinary barriers and repudiating the importance of the issues we have studied up to now? Over the next few decades many of our techniques are likely to become much more adequate, but what is our role until then? It seems to me that we belong to one of the institutions of society whose members and operating styles need to be shaken up quickly—we need to have our awarenesses of reality enlarged and refined and revised if we are to make our contribution in good conscience and with significant effect.

One might decide, of course, that even with all these conflicts and changes and even without the participation of the scholars, some kind of accommodation will be worked out. Probably so, but there is the possibility that the accommodation will be one we will not like. And there is also the possibility that there will be no accommodation. Certainly ours would not be the first society that disappeared because it could not find a way to accommodate in time to changes generated within it by its own momentum and style.

DISCUSSION

HAS THE HUMAN CONDITION CHANGED?

Question: What Mr. Michael has done is to restate what the human condition is, and always has been, put of course in the context of problems of our own time. The novelty of the situation escapes me. At the same time I grant that he has given us an agenda of problems for contemporary society. My impression that the situation is not really novel began when Mr. Michael gave the Industrial Revolution of the eighteenth century as an earlier example of a technological revolution that had produced problems because of the rapidity of the rate of change. The two problems he gave as examples in this case were the emergence of the urban slum and the breakdown of a world view. Many of the medieval cities had something analogous to slum problems long before the Industrial Revolution. Similarly the breakdown of a coherent world view occurred long prior to the technological change that characterizes the eighteenth century Industrial Revolution. I wonder, therefore, whether we are here placing the responsibility on technology for things that really reflect a much more general human condition.

The problems produced by a changing military technology for example, are nothing new. The religious wars of 1618-1648 saw the population of Germany reduced from some eighteen million to about four million. The casualty rate for that type of war was far higher than it is in modern warfare. The concern in the Renaissance about the change in military technology has its echoes today.

It is worth noting that even in the time of Plato it was recognized that science—in this case the science of persuasion —could not be neutral.

Mr. Michael: I have great difficulty understanding how you fail to see the profound differences between your examples and what I have been talking about. For example, cities before the Industrial Revolution were quite different social entities than the industrial city with a population of factory workers and displaced rural workers. You just can't say of the two cases that the crowded conditions of city poor, as such, have analogous social meanings for the city or for the rest of society. Furthermore, I am asserting that when you can destroy a civilization as large and complex as ours, or when you can smash up the Van Allen radiation belts for thirty years, or can alter the genetic characteristics of generations to come, or can accidently put DDT in the livers of fish in the oceans of the world as a result of spraying crops at home, or when you can, not abstractly and hypothetically, but actually, manipulate the public on a grand scale, you get differences in degree that, I feel, amount to differences in kind.

Question: But did not the invasion of the Huns lead to the destruction of an entire civilization?

Mr. Michael: I don't know of any methodology or criteria that would tell us here tonight the degree to which such statements as "The Huns destroyed an entire civilization," and "Nuclear weapons could destroy an entire civilization," can be compared meaningfully, much less how one could extrapolate implications for the present from them. Regardless of what the Huns did to the world, we have *new* capacities for doing enormous things to the world, and today people live in different ways and in different numbers than ever before. It is with the people living now and in the future in our kind of world, with our kind of capacity to create or destroy ourselves that I am trying to deal.

Comment: In further support of Mr. Michael's position, the argument we have just heard would lead one to think that the Greeks invented nuclear physics and that there is nothing radically new to nuclear physics.

Mr. Michael: To continue, it is true that many dinosaurs had very great problems accommodating to their environment. That might be called the dinosaur's condition. Problems arose all the time, and I can imagine someone pointing to a coming what-ever-it-was and being answered in the spirit of "Oh, we have always had problems." But the essential fact is that the dinosaur is gone.

Question: What criteria could we develop to measure objectively that the situation is more precarious today, when we are facing extinction, than it was in the seventeenth century, when they could cope with their problems? This is obviously a very difficult question; I do not know what criteria might be used other than one man's appeal to historical rhetoric and another's appeal to contemporary rhetoric.

Comment: Perhaps this issue can be narrowed. Mr. Michael stressed the fact that one aspect of contemporary technological change is overwhelmingly negative in its essential import for humanity. To this he added a subthesis, namely, that there is a concentration of power, which heightens this negativism to such an extent that it makes a qualitative change in mankind's situation.

Obviously our overriding concern today is the question of nuclear war. But a far from minor consequence of technological change is, for example, the automobile with which we murder some forty thousand people each year in America alone.

But it is possible to see these same developments from a positive point of view. I would say that though it may be true that we are closer to total extinction than we have ever been, it is also conceivable that we are closer to gaining control of large-scale wars than we have ever been.

Also, the same instrument, for example, the automobile, that leads to a very large number of deaths each year also has a life-saving, life-enhancing attribute. Modern technology cuts two ways, and our first commentator was merely saying that he does not really see how we can sort out the overwhelmingly negative implications of technological change from the positive implications in order to state clearly whether the human situation has altered in a unique fashion for the worse due to these technological developments.

Mr. Michael: To continue with the analogy of the two-edged sword—what I have tried to emphasize is that the technologies of today are very much a two-edged sword, and that the sword cuts much deeper than any sword that has ever existed, both in its positive and in its negative aspects. My concern is that we be prepared to deal with the negative aspects in order to be able to reap the benefits.

OTHER CRITICAL PROBLEMS: LEISURE, POLICE POWER,
UNLIMITED POWER FOR DESTRUCTION AND PRODUCTION

Question: Mr. Michael has understated his case. He did talk about leisure, but unfortunately he did not indicate that the elite groups of society, particularly the intellectuals, are going to have to work harder in the future than they have ever done in the past. A very difficult problem is apt to arise when those who work hard in leadership positions attempt to persuade those elements in society who do not have much work to perform that it is all right to be relatively idle.

Mr. Michael has also raised the general question of the invasion of privacy that has accompanied recent technological change, using the Job Corps as an example. I would suggest that invasions of privacy by the police is a much more serious development.

It is my feeling that what is truly new in our time is the fact that man has either achieved, or now has the ability to achieve, unlimited power over his environment in several dimensions. Perhaps the most obvious today is the drive toward unlimited destructive power, but that is paralleled by the drive toward effectively unlimited productive power. In historical terms the shift from the industrial age to the cybernetic age is as big a shift as the shift from the agricultural age to the industrial age; the major difference between the two shifts being that the former is taking place over a generation or even less while the latter took centuries to work out.

Two problems, in particular, emerge out of this extremely rapid change. The first is the general problem of coping with such rapid rates of change. The second problem is the necessity to control unlimited power, something man has never really been forced to do. I do think we have an opportunity to create an infinitely better society than man has known in the past, but this is only possible if we can change our social system, particularly the system of nation states, without war. We must make these changes in our social and political systems without war or revolution because both threaten to unleash unlimited destructive power.

The problem then becomes one of developing a leadership able to bring about the necessary changes. Effective points of leadership are extraordinarily difficult to find today. People who might be leaders instead operate according to the consensus. One can say to them, "But you know this is not true." They will answer, "Yes, but it is not convenient for us to state that it is not true." We are not going to be able to make noticeable progress until, for example, the computer companies state very clearly what they already know about the consequences of computers, and the generals state very clearly what they know about the nature of modern warfare. I see little evidence that our society is developing the form of leadership it is going to need in order to cope successfully with the speed of change today.

Mr. Michael: I agree entirely. The ability to make constructive use of the new technologies, as well as the ability to avoid the very great risks of misuse, requires a very high degree of wisdom on the part of policy makers and leaders. We have now no way of creating such wise people in any systematic or unsystematic fashions—and, what is more important, we seem unwilling to alter institutional arrangements in such a way they will encourage and respond to wise counsel.

Question: The direction of technological change is obviously controlled by various institutions within our society. How can we change institutions fast enough to deal with technological change? Some institutions tend to be very conservative. Those institutions that today are of crucial importance in determining the rate of technological change and its direction are, in some cases, just such conservative institutions. The military, of course, is the prime example. But the university also comes to mind.

How do we deal with this paradox of rapid technological change taking place under the aegis of institutions which themselves do not easily change? Is it possible that it is just as well that these institutions do not tend to change rapidly?

Mr. Michael: An area for social research I believe to be very important is, How and why do institutions change and how can they be made to change quickly? We know all too little about the character of our major social institutions.

Question: Mr. Michael is primarily concerned with the tremendous magnitude of recent technological developments and believes that their nature and size make them different in kind from what has previously happened. It may be that the fact that these developments are concurrent also adds to the gravity of the problem. But, on the other hand, the very concurrent nature of these developments offers opportunities for solutions that would not otherwise be available. For example, our new capacity to engage in social engineering may make it possible to handle the cybernation problem.

Mr. Michael: This requires overcoming enormous theoretical and operational problems of phasing and making sure that these developments occur in the right sequence so that the outcome is felicitous. Underlying everything is the problem of who makes the decisions about which potent techniques are to be used in what way and for what purpose.

Comment: Our society is one in which the level of aspiration tends to be fairly low and the level of projection into the future quite short. There is, therefore, an implicit advantage for the kind of proposal made by Robert Theobald—deliberately utopian in character, I should say—which simply compels people to think about the possibilities of the future.

Mr. Michael: I should agree. One of the major requirements for coping with the future is to have more people speak and write in order to alert the public to the need to change our institutions in time to prepare for the future.

If we are going to change institutions deliberately so as to produce some of the changes in the social environment that seem desirable in order to take advantage of new technologies, we are going to require a much greater ability to manipulate people than we have now. A greater ability to manipulate is one of the positive fruits of modern technologies, but the negative aspects take on new meaning because of the potency of the techniques available. Who is to make the decisions on these policies? What is to be the individual's role as a citizen?

THE IMPLICATIONS OF CONSPIRATORIAL STRUCTURES

Question: Mr. Michael implies that there is some sort of conspiratorial element that will use the power of these new technologies against the society of which it is a part.

Mr. Michael: First of all, at any given time decisions are made within significant power centers of our social system on how to present information for political purposes and to the advantage of the power concerned. This may involve, if you will, a kind of ad lib. "conspiracy." For example, many believe that such a "conspiracy" existed between the AEC, the CIA, and SAC in matters having to do with the information and policy made public about levels of nuclear radiation expected and tolerable in a nuclear war. Some competent people feel that a number of government offices "cooperated" with offices in the Department of Agriculture to obscure or refute the contention that pesticides had been misused under USDA policies. At any given time certain coalitions have access to certain information—and I think this will increasingly be the case— to which the general public does not have access. Thus the coalitions can select from this information what the public is to know, rather than have the public respond to the significant factors of the reality.

Question: Do you really think that the scientists in the Department of Agriculture are convinced of the evils of pesticides and do not tell the public?

Mr. Michael: Not in any simple sense. These scientists are convinced that what they are doing is right—and "right" can mean everything from, "there's no proof to the contrary," to "our only responsibility is to cover up this blunder for the sake of the survival of the agency." Probably they are convinced of their rightness in just the way that the Federal Radiation Council was convinced, that adjusting the "safe" level of radiation up as the amount of fallout increased, was in the national interest.

Question: What you are arguing for, then, is more information, more widespread information, and a greater sense of responsibility on the part of those people who have the knowledge.

Mr. Michael: Yes. And much more effort should be made to determine what information is useful and how to assign responsibility to use the information—and make it stick.

Question: We are living in an age of greater uncertainty than mankind has ever experienced before. Much of this uncertainty is due to technology. What bothers me about the position of people such as Mr. Michael or Mr. Theobald is that there seems to be implicit in their approach a demand for certainty and order. I think that they are at heart utopian; I think that they believe that if they were able to get their hands on the levers, life would be better. Implicit in the position of Mr. Michael and Mr. Theobald is that there is one best way, one best universe.

However I see the situation in another way. There may be more uncertainty in the world, but this is another way of saying that there is more opportunity, more opportunity for joy, life, agony, death. I am optimistic in that I think that there is more individual consciousness today. This is reflected in the fact that groups in society do get together, they do have the kind of discussion we have here, and they do substitute consensus for war and revolution.

Mr. Michael: I categorically reject this formulation of my position. Everything I've said was intended to emphasize that we do not now understand, nor are we systematically trying to, what alternatives are or will be available to us. I couldn't possibly argue that there is "only one best way"; I am making a plea that we learn enough to avoid the disasters that will keep us from finding and enjoying the felicitous alternatives—and that we learn enough so that men *can* recognize the alternatives available to them in such an overwhelmingly complex world. The challenge is how to deal with this new order of complexity and uncertainty and make it productive rather than destructive.

The Need for Technological Change

by HENRY H. VILLARD

Professor of Economics, The City University of New York

DONALD MICHAEL suggested that he was a pessimist, and that we faced some rather "threatening" problems in adjusting to the technological change that is likely to occur over the next couple of decades. Actually, from my point of view, he is really quite an optimist. For, as I see it, the problem is not merely to adjust to prospective technological change over a couple of decades, but, rather, to achieve a much more rapid rate of technological change than we have yet achieved, to make the adjustments that this more rapid change involves, and to do so not for a couple of decades but for a couple of centuries!

Let me make clear that I did not take this assignment because I enjoy speculating about the future. In fact, I was more than a little diffident about accepting it because of the low opinion we all hold of such speculations. But the difference between Donald Michael and myself, as I see it, simply reflects the fact that as a group we have vastly different ideas about whether continued rapid technological change is or is not desirable. Some of us seem to feel that science and technology, if they have not already solved the production problem, can be counted on to do so in the near future, so that all that remains is to adjust to abundance—in other words, to make sure that we are not swamped by the goods pouring out of our automated factories in the way in which the sorcerer's apprentice was swamped by the water carried by his automated broomstick! At the other extreme, others of us—myself included—feel that rapid technological change continues to be urgently needed, perhaps more urgently than ever before.

The only way I can see to decide between these diametrically opposed views and develop some consensus regarding the need for technological change, or the desirability of the research that makes such change possible, is to try to reach some agreement on the amount of additional production that will be needed. Obviously much depends on what we mean by "needed." Specifically, in his talk, Professor Sebrell suggested that the world is not likely to run out of food—at least not before something else intervenes to halt population growth. I suspect many of us took more comfort from this than we should have. For Professor George Stigler has demonstrated that before World War II all essential dietary requirements could have been met at a cost of around forty dollars a year—or almost certainly less than a hundred dollars at the present time. It follows that 3 or 4 percent of what the average American today spends for all the goods and services he consumes is probably enough to provide him with his essential dietary requirements. All that Professor Sebrell said is that meeting the essential dietary requirements of even a very much larger world population should not, as a technical problem, be particularly difficult—and I entirely agree.

But is this *all* we want to provide? Even Professor Stigler did not suggest that anyone should *live* on his diet—which, as I recall it, consisted exclusively of beef liver, pancake flour, and spinach! In short, there is an immense difference between including in "need" only enough production to meet the essential dietary requirements of the human race, and considering necessary some of the things on which the average American spends 96 percent of his income. I must admit that I do not see how we can avoid including a lot more than dietary essentials, but the minute we do, the problems we face become nothing less than immense.

The estimates, which follow, are quite crude: I have felt it more important to discuss their broad significance than to devote time to qualifications and refinements. I propose to start with the United States. If there is anything wrong with this seminar, it is that it has confined itself almost exclusively to the 6 percent of the world's population that lives in the United States. But let me briefly continue our provincial tradition and consider what is involved in providing everyone in the United States in the year 2000 with $20,000—which is, on the average, the salary now received by professors in our better universities.[1] In 1950 the average American worker received somewhat less than $5,000; between 1950 and 2000 our population will about double in size. Thus, to bring everyone up to an average of $20,000, total production would have to increase eightfold between 1950 and 2000. This in turn means that total real income would have to increase at more than 4 percent a year over the period, or significantly more rapidly than the average increase since 1875. Even more important, income per worker would have to increase fourfold over the period, which means that the rate of growth of income per worker would have to average 3 percent, or be about 50 percent higher than the average achieved since 1875. In fact, if the historic 2 percent rate continues, the average American will not receive $20,000 until 2020—or 55 years from now. In short, if I read the lesson of these simple statistics correctly, they suggest that anyone who believes that the production problem has been solved believes that the average American should not, within his lifetime, aspire to live on the same level as a better-paid college professor does today.

But I am not primarily interested in college professors, or

[1] All statistics cited are based on purchasing power in terms of 1960 prices.

even in the United States. Rather, I want to consider what is involved, not in merely feeding the human race, which was Professor Sebrell's topic, but in providing the human race with some appropriate level of living. Let me temporarily postpone the obvious question as to what I mean by "appropriate" and, without debating the matter, start by illustrating what is involved in raising everyone *to half the average level of living that we had achieved in 1950.*

In that year, in the world as a whole, average income per *person*—rather than per *worker*—was a bit over $300, while ours was about $1,800, so that a *threefold* increase in world-wide production would have brought everyone up to half our level—provided, of course, that *all* the increases, including the increase in our own production, were to be divided among those outside North America and Oceania (who are now at half the U. S. level). If a threefold increase seems relatively small, it is because European incomes average about one third our level, so that the increase in *total* production required to bring the world up to half our level is not particularly spectacular. But the increase that would be needed in particular areas is much greater. In Asia and Africa, for example, which contain between them almost two thirds of the world's population, nominal income per person in 1950 was under $100, so that a *ninefold* increase in the production of the two areas would be needed to bring average incomes up to half our level. Let me repeat at this point my warning that these estimates are crude, being based on unadjusted statistics from the UN, but the broad picture is not likely to be greatly changed by refinements.

This, then, is the dimension of the static problem: for the 93 percent of the world's population who live outside North America or Oceania, a *threefold* increase in the world's production would raise their living levels to *half* our 1950 level, provided that all the increase went to those below that level. But for the two thirds of the world's population who live in Africa and Asia, if they had to depend on increases in their own production, a *ninefold* expansion would be necessary.

One thing we can be sure of, however, is that the world is not static. Suppose we take a look at what is involved in raising world-wide living levels to half of the level we will probably have in 2000. Between 1950 and 1960 the rate of increase of the world's population was more or less constant at 1.5 percent per year, but from 1.6 percent in 1960 the rate of increase has risen .1 percent each year, and has now reached 2.1 percent, which means that our numbers will more than double by 2000. In other words, had the 1950 rate continued, there would have been 5 billion people in 2000, but it now looks as if there will be 6.5 billions in that year—or 30 percent more people. Moreover, the utterly unprecedented decline in infant mortality since World War II caused an unprecedentedly large number of children in underdeveloped areas to survive, so that five to ten years hence there will be an immense increase in the number of those of prime child-bearing age in these areas, which in turn means that there will be a sharp increase in the rate of population growth *even if birth and death rates do not change.*[2]

Actually the death rate is still falling: it may well be that, after atomic energy, DDT will turn out to be the most significant invention of our century! Admittedly, on the side of the birth rate, some new things have been added—the pill and the coil. Unfortunately the former, though simple, is expensive, and the latter, though cheap, requires medical personnel for insertion. The long-range prospect for population growth is a matter to which I want to return. But, given the age distribution in underdeveloped countries, the likely decline of the death rate, and the difficulties we face in bringing down the birth rate with

[2] To illustrate: in England and Wales in 1871—which was the year in which the ratio reached its peak—the number of those between 0 and 9 years of age was 1.73 times as large as those between 25 and 34. The same ratio today for Thailand and Indonesia is 2.15, for Mexico 2.43, for Pakistan 2.47, and for the Philippines 2.60!

the birth control techniques now available, the most optimistic assumption I can make is that the increase in population over the next 35 years will average out at 2 percent, or about the present rate, which (as already mentioned) will give us a world population of 6.5 billions in 2000.

If this is correct, it follows that our problem is to raise 6.2 billion non-Americans in the year 2000 to half the level that 300 million Americans will have in that year. I estimate that in 2000 half our level is likely to be $2,500 per person, assuming continuance of the 2 percent increase per person that has prevailed since 1875.[3] For America, as we are merely projecting the past rate of increase, such a level can obviously be reached. But for the rest of the world the required increase in production would be from under $700 billions to over $15 trillions—a rise of more than twenty-two times the 1950 level. Even if we include our own production in the 1950 base, total world production would have to increase sixteenfold to provide all Americans with an average income of $5,000 and everyone else in the world with an average income of $2,500.

Perhaps I should quit while I'm ahead: 2000 is a year some of us will see and twenty-twofold increases are comprehensible. But let me speculate briefly about the twenty-first century. For, just as Professor Sebrell's talk implied that all that was needed was to feed people, so many who worry about the population explosion seem to feel that all that is needed is to bring about a decrease in the birth rate. Even with nothing more than the coil and the pill it ought to be relatively easy to bring about *some* decrease in birth rate on a world-wide basis, as there must be literally hundreds of millions of unwanted babies born every year. But the real problem we face is not merely slowing population growth, but *halting it*. Just how soon cessation must be achieved depends on the sort of world we want hereafter. In

[3] The estimates continue to be based on purchasing power in terms of 1960 prices.

1950 the entire land area of the world (including Greenland, Antarctica, and the Sahara desert) came to 650,000 square feet per person. The average American had 500,000 square feet of unusually fertile land, Englishmen had 32,000 square feet, and New Yorkers 1,300 square feet. Geometric progressions are impressive: exactly nine doublings of the world's population—which at the present rate of population growth will require 315 years—will cause the land area of the world to be as densely populated as New York City. Now I am a native New Yorker who loves the city, but I freely confess that a world consisting exclusively of New Yorks does not strike me as ideal; from this I deduce that halting world population growth before nine doublings occur has decided merit.

But halting growth will take a bit of doing, because families will have to be induced to confine themselves, in effect, to 2 children. Beyond the basic two (who replace the parents), the average family may have a fraction of a child to compensate for (1) women who never marry, (2) infertile couples, and (3) girl babies who die before they have their two babies. The Population Reference Bureau estimates that at the present death rate in the United States, a stationary population requires that the average woman have no more than 2.2 children; were she to have as many as three, the population would increase by about 50 percent in each generation.

For what it's worth, my guess is that it will be impossible to induce women to confine themselves to 2.2 children voluntarily. The number of cats and dogs in the United States has increased immensely in recent years; as incomes rise I suspect that many families in the future may well turn to children, who are, after all, more fun. But if I am right that women will not restrict themselves to 2.2 children voluntarily, then ending population growth will take time, since we are a long way from being ready to use compulsion. I believe, therefore, that it will be a considerable achievement to bring about cessation of population growth before 2100 and with less than two further doublings (beyond the doubling that is likely between now and 2000). If this guess proves correct, then stability would come with a world population of 25 billions—ten times the population of 1950. But an exact estimate is obviously not important. All I really want to do in this brief excursion into the twenty-first century is to point out that achievement of a static population involves acceptance of an average family that is very small by present standards and that, until acceptance is actually attained (with or without compulsion), population growth alone may require *two further doublings of world-wide production even without allowing for any increases whatsoever in living levels in the twenty-first century*. So the need for increases in production will not end in 2000.

Now, assuming I have not lost too many of you in the twenty-first century, we shall go back to 2000. Let me remind you that our conclusion was that it would take a sixteenfold increase in the total world production of 1950 to provide all Americans in the year 2000 with an average income of $5,000 per person and everyone else in the world with an average income of $2,500. What would such an increase involve? Not much from a purely statistical point of view: a sixteenfold increase involves four doublings, which would take just under 75 years if a 4 percent rate of increase in real income could be maintained. In other words, if world-wide income could be increased at a rate only a bit higher than real income has been increasing in the United States since 1875, our proposed levels could be attained—not, admittedly, in 2000 but by 2025. This conclusion disregards the major shifts from "haves" to "have nots" that the overall figures conceal—and at a time when incomes are rising more rapidly in developed than in underdeveloped areas.

But suppose we look at what is behind the pure statistics and see what they imply in terms of natural resources. Let us assume that increased efficiency causes resource consumption by everyone in the world, when their income averages $2,500, to be no greater than our consumption was in 1950, when we had incomes which averaged $1,800, and that an increase from $1,800 to $5,000 in our average income will merely double resource use. We estimated that there would be 6.2 billion non-Americans in 2000, or 41 times the 150 million Americans alive in 1950, and that there would be twice as many Americans consuming twice as many resources in that year. It follows that resource use in 2000 will be 45 times our rate of use in 1950, if the proposed income levels are achieved and the assumptions just made prove correct. Even if no resources whatsoever are used between 1950 and 2000, so that all present supplies remain available, then, starting in 2000, the predicted rate of use would exhaust known reserves of copper in less than three years, and known reserves of iron ore and petroleum in less than twenty-five years; only coal would remain relatively plentiful, with reserves equal to, perhaps, three hundred years of use.[4]

I am well aware of the kindly chap who worried more than 100 years ago as to whether the inhabitants of the twentieth century would be able to read at night because of the dwindling supplies of sperm oil. Let me stress, as strongly as I can, that I do not present these estimates to suggest that *specific shortages* of natural resources are likely to be important; there is no doubt whatsoever that we can replace fossil fuels with either atomic or solar energy, and steel and copper with aluminum, magnesium, or some plastic. But the shift to other energy

---

[4] These estimates assume, in the case of coal, that all coal found in seams more than one foot thick down to 4,000 feet, and in seams more than two feet thick between 4,000 and 6,000 feet will be utilized; in the case of petroleum, that fourteen times as much oil will be produced in the future than has so far been recovered; that natural gas will provide an additional amount equivalent to 40 percent of the liquid petroleum total, and that petroleum extracted from shale will provide as much again as liquid petroleum. On the other hand, my assumption that resource use per dollar of income 50 years hence will be at two thirds our 1950 level may be overly pessimistic if technological change in the interim is able to achieve rapid conservation in resource use.

sources and to other metals or metal equivalents will not necessarily be easy. Most of our present stock of capital goods has
been developed to exploit fossil fuels, and steel remains our
outstandingly important metal. It does not, therefore, seem to
me to be an exaggeration to say that significant economic development on a world-wide scale involves nothing less than a
change in the entire technological base on which our civilization presently rests.

To what conclusions do all these estimates lead? Not surpris-
ingly, they suggest that the urgency of the need for technological change depends upon what we want to achieve. Specifically, as we have seen, a significantly more rapid rate of
technological change than has been realized since 1875 will be
needed, if the average American is to attain an annual income
of $20,000 by 2000, but maintenance of the past rate will suffice
if we can wait until 2020. Now I do not know whether the
average American *should* aspire to $20,000; still less am I clear
whether the average inhabitant of the world *should* aspire to
live half as well as an American. In fact, let me confess that
who owes what to whom confuses me. Most of us, I am sure,
feel an obligation to devote resources to raising the living level
of Negroes in the United States. But what do we—and I include
American Negroes—owe Negroes of presumably equal potential who are at a very much lower level of living in the Congo?
These are obviously matters we could debate endlessly. But
one thing I think we can say with some certainty. If we accept
raising the population of the world to any significant fraction
of our level of living as our task, then it is obvious that, far
from having solved the production problem, science has not
begun to meet the challenge that we face—has not, so to speak,
begun to fight. For bringing the 6 billion or so people who are
likely to be living on our planet in 2000 up to half our level
involves, as we have seen, a sixteenfold increase in total production or a twenty-twofold increase in production outside the

United States. And these estimates do not make any allowance for continued increases in population—or living levels—in the twenty-first century. In short, if the assignment is to concern ourselves with world-wide living levels, then the real danger is that we will cease to encourage scientific research and technological change just when they are most urgently needed, because many others beside the Center for the Study of Democratic Institutions fear that the magic broomstick has gotten out of hand.

Suppose we accept the need to concern ourselves with world-wide living levels; what should we do? Let me warn you, there is no consensus among economists. Edward F. Denison, in his recent study for the CED on *The Sources of Economic Growth in the United States,* estimates that 47 percent of the increase in output per man-hour between 1929 and 1957 was caused by improvements in the quality of our labor input (as a result of both reductions in hours and increases in education), 27 percent resulted from technological change, 16 percent from economies of scale, and 10 percent from the net impact of all other factors. Having made it clear that Denison's study is by far the most careful yet made, I have no hesitancy in characterizing his overall results as nonsense! I just do not believe, as his results imply, that, using the capital equipment and technology that were available in 1929 and producing only the products that were then in existence, the 1957 labor force was so superior that it could have turned out almost 60 percent more than was produced by the 1929 labor force! In my, admittedly undocumented judgment, somewhere between two thirds and three quarters of what has been achieved since 1929 has been the result of technological change, broadly defined. This is not to deny that education, economies of scale, and other factors have played a role in the increases we have achieved; the point simply is that new technology has, in my judgment, been of overwhelming importance.

Assuming I am right, what specifically do I recommend we should do? First of all, I am reasonably certain that we are not going to raise world-wide living levels appreciably without halting population growth. Since independence, total production in India has been increasing at about 3.5 percent a year —roughly at the same rate that our own production has been growing since 1875. At the start, Indian population was increasing at only 1 percent a year, so that the annual rise in per capita income was 2.5 percent—or an increase of about $1.50 in a per capita income of $60. But, with the rate of increase in total production unchanged, DDT and better diets have recently lowered the death rate to the point at which the population is increasing at close to 2.5 percent a year—so that per capita income is now rising at little more than 1 percent a year —which involves an annual increase of about 70 cents! In short, unless there is a drastic decline in Indian fertility, the second half of the twentieth century is likely to increase average Indian incomes from $60 to between $80 and $100, and the Indian population from 350 millions to between 900 millions and a billion. I rather doubt if Indians will consider this a wholly satisfactory achievement.

One thing that I suspect is going to increase Indian dissatisfaction is that it looks as if during that period, "To him that hath will be given, and from him that hath not will [relatively at least] be taken away." My estimate was, that between 1950 and 2000 the income per person in the United States is likely to increase from $1,800 to $5,000—an increase of 175 percent. But I have just suggested that Indian income over the same fifty years is likely to increase by, perhaps, 50 percent. This in turn means that Indians, who were at, perhaps, one thirtieth of our level when they attained independence and started intensive development, are probably going to find themselves at one fiftieth of our level after a half century of hard work.

Need we concern ourselves with what happens in India? Are

we our brothers' keeper? One way of offering some objective evidence on the matter is to compare the world distribution of income with income distribution in particular countries. At present, the 6.5 percent of the world's population that lives in North America receives, perhaps, 36 percent of the world's production. Now the situation has clearly been worse in some countries in the past: Professor Simon Kuznets has estimated that 5 percent of the population of Great Britain received 46 percent of the national income in 1880. But I have the impression that this distribution was not considered entirely satisfactory.

Moreover, it has changed drastically—largely for reasons that are by no means fully understood. Specifically, by 1947 the top 5 percent in Great Britain were receiving only 24 percent—approximately half their share in 1880. And in the United States, according to the Department of Commerce, the top 5 percent received only 20 percent of the national income in 1959. Nor is this all. For out of the 20 percent came savings and highly progressive income taxes, so that I doubt if the top 5 percent of income recipients account at present for more than 12 percent of consumption. In short, while foreign aid and foreign investment mitigate the international concentration to some small extent, it appears that the 6.5 percent of the world's population living in North America receive roughly three times as much of the world's consumption as the top 5 percent of American income recipients receive of our consumption. And this concentration is steadily increasing as the developed areas grow more rapidly than underdeveloped areas.

How long can such a situation continue in a world where atomic capacity is becoming increasingly general? I, of course, have no precise answer. But I do know that the 1880 concentration in Great Britain had much to do with the growth of socialism—and the sharp decline in concentration since then had much to do with the relatively limited spread of communism. For Marx's prediction, that the rich would get richer, and the poor, poorer, was completely wrong for the countries of the Western world. But for the world as a whole his prediction is now coming true! Thus, though our most immediate problem may be not to blow ourselves up in a nuclear war, I suggest that bringing population growth to a halt is next in importance. In fact, I am not sure but that, in the last analysis, both are aspects of the same problem. For, continued avoidance of atomic war seems to me likely only if we so manage our affairs as to move steadily toward viable standards of living for the entire human race.

What else needs to be done? While population growth is being brought to a halt, I suggest that we should systematically encourage research at all levels. In the United States not much more than .5 percent of the national income is spent on basic and applied civilian research taken together. Even more important, industries accounting for 60 percent of the national income spend only a small fraction of 1 percent of their income on research, and are responsible for only around 5 percent of all spending on research. In sharp contrast, the remaining 30 percent of the civilian economy spends close to 5 percent of its income on research and is responsible for perhaps 45 percent of all research. Finally, the military sector, though it accounts for only 10 percent of the national income, spends about 15 percent of its income on research and is responsible for about half of all research. Obviously the primary problem is to devise ways of encouraging research in the 60 percent of the economy that is presently responsible for only 5 percent of all current research.

Finally, as those working in our Point Four program have learned, the technology of developed countries usually has to be extensively changed before it can be used to meet the needs of underdeveloped areas. If we are serious about promoting world-wide economic development, and do not think of our

A.I.D. program as charity for immediate political advantage, then it seems to me that systematic and continuing research to create technology appropriate to developing countries may well turn out to be by far the most important contribution that we can make—far more important than the probable contribution of the general funds we are likely to provide.

In short, as I see it, it is vitally important to increase our rate of technological change—for two reasons. First, because it is of the essence in helping the underdeveloped countries to develop; and, second, because rapid growth in the United States will contribute indirectly to the solution of world problems by increasing our willingness to offer the aid to underdeveloped countries, which seems to me essential if we are going to create a viable way of life for the human race.

But let me end on a note of caution, so as not to leave you with false optimism. To achieve by public policy even a small change in the rate of growth of total production will be difficult in the United States—and probably even more difficult in most developing areas. In fact, increases from the historic 3.5 percent rate of increase in total production to 4 or 5 percent in either India or the United States would be major achievements. So I am afraid that scarcity and, therefore, the need for technological change (and, I might add, for the science of economics), are likely to be with us for a long time to come!

Discussion

THE POPULATION PROBLEM: OTHER FACETS

Question: What studies have been made to discover where population growth has decreased *without* technological development—or *without* modern contraceptives?

Mr. Villard: Obviously some peoples have been able to restrict their population growth without the availability of modern birth control techniques. The prime example is, of course, Ireland. Their solution to the population problem has been based primarily on postponed marriage. Postponement started, however, with the potato famine, which cut the Irish population in half. If that kind of catastrophe occurs, I am certain a radical change takes place in the number of children people have, even without modern birth control techniques. But, certainly, this is not likely to be a common, nor a desirable, way to solve the problem.

Question: I think the phenomenon of population reduction is more common than you imply. The population of the ancient Mediterranean world suffered radical decreases. There have been tremendous reductions of population all over the globe throughout history. My point is that you would have to study these cases before making any firm predictions about population. You might discover techniques, which would help solve the population problem.

Mr. Villard: I completely agree that the problem is not merely a question of developing birth control techniques. Rather, it involves the creation of a social environment in which people want to limit births. If the motivation is strong enough, other peoples will do what the Irish did. But I know of no underdeveloped areas today where there has been any significant population limitation based on the use of crude techniques.

Comment: The technology of birth control is actually moving very rapidly right now.

Mr. Villard: I wish I could agree with you. The oral contraceptive is still expensive, hard to get, and effective only for relatively sophisticated people. As far as I know, neither the pill nor the coil has had any effect on any area of any size— though, of course, the situation may be different in the future.

Question: If we assume further development of birth control techniques, which I think is a realistic assumption, do you think that your population estimates are realistic?

Mr. Villard: But even with cheap and effective birth control techniques, population growth is likely to continue because of *wanted* children. The American population, for example, is today expanding rapidly, principally because of wanted children. And it may well continue to expand at a rate not very different from the present rate (almost certainly over 1 percent per year) because of wanted children. This means that, as we move toward bringing population growth to a *halt,* the problem becomes increasingly difficult. We shall have to make people want much smaller families than they now have, at a time when their rising standard of living is going to make the cost of raising children relatively much less important. That, fundamentally, is why I believe the population will continue to increase well into the twenty-first century.

ALTERNATE SOLUTIONS TO THE PROBLEM OF POPULATION INCREASE
AND STANDARDS OF LIVING

Question: Mr. Villard in substance has presented two solutions to the interlinked problem of population and standards of living. First of all, given continuing increases in productivity, he looks for some means of limiting population growth. Secondly, he looks for some means of redistributing income internationally, which is essentially a political problem. I would suggest that there is a third solution—the redistribution of the population of the world. A major trend, historically, has been a movement of a good part of the world's population in the direction of areas of higher productivity.

Mr. Villard: My reservation about this as an immediate answer is that it would reduce the pressure to slow down the rate of population growth and, therefore, postpone the ultimate solution.

Let me emphasize the acuteness of the present population problem in certain areas. Take, for example, the little island of Mauritius. Today it has an average population density equal to that of Great Britain—about six hundred people to the square mile—and its population is presently doubling every twenty-five years. Mauritius is going to solve its population problem soon, because it has no way to avoid facing it squarely. In fact, a monsignor recently told me that he had heard the Bishop of Mauritius talk for more than an hour about the urgency of the Mauritian population problem. As I see it, in areas such as Mauritius, the problem is likely to become so acute that the society will break through the various obstacles that prevent the adoption of a birth control program. I would hate to see emigration postpone such a solution.

Question: Could we explore more carefully the relationship between what you might call economic pressures and the motivation of a society to change. There are three separate facets to the situation. There are economic pressures, there are motivations to change, and, finally, there is the capacity to do something about the problem. Mr. Villard implies that we are going to get a solution to the population explosion only when everybody is concerned about it. Perhaps the only real way to solve the world's population problem is to create such a population crisis in the United States itself that the scientific and technological intelligence and capacity of the United States will focus on the problem, solve it for the United States, and thereby solve it for the rest of the world.

It is not at all obvious, however, that even if all the people in a society get very concerned about the problem—like your Bishop of Mauritius—they have the capacity to do anything about it.

Mr. Villard: Let me say, first, that one of the major factors you overlook is that the present population density of the United States is not much over ninety people per square mile, while it is over six hundred per square mile in Mauritius. Indeed, an authority such as Edward F. Denison (and I am inclined to agree), argues that we still benefit more than proportionately from population growth. I do not, therefore, think that the Mauritians are going to insist that we in America practice what we preach before they themselves start a program for population control. What is important in the Mauritian situation is that one third of the population is Catholic, and two thirds are Hindu. The population problem is not going to be solved there until both the Hindus and the Catholics are prepared jointly to do something about it.

But I do agree with you that we need not only a better birth control technology, but that we also need to know much more about how to get people to use the techniques we do have. The Ford Foundation has been working very consciously on both fronts. The current study in Taiwan is an example of a program, which is aimed at determining, by a controlled series of experiments, how best to reduce the birth rate with presently available techniques.

Question: Mr. Villard has emphasized the need for technological progress in general, but it seems to me that what is really required is applied research in the technological needs of developing countries. For example, the International Harvester Corporation will not try to develop agricultural implements for developing countries unless it is offered special incentives to do so. I believe that we have to develop very specific goals for applied research, yet I wonder how we can achieve this in a free society. How can we persuade a business that can secure very high profits by developing the technology in the United States further, to think, instead, in very broad international terms?

Mr. Villard: I do not know how you develop a basic frame of mind, except, perhaps, by having sessions like this. But I do believe that, to get people to engage in applied research oriented to the technological requirements of developing countries, we must set up major research stations in various regions of the world. To some extent steps have been taken in this direction. For example, the Rice Research Institute, supported by the Ford Foundation, has done a magnificent job in developing techniques to increase the yield of rice. This type of research institute can, in my opinion, be very useful, since it can be directed toward the particular technological needs of an area. Let me repeat my conviction that the establishment of such centers would typically pay off far more than other types of foreign assistance programs.

Of course I agree with you that it is relatively more urgent to improve the technology of the underdeveloped areas than it is to achieve technological improvements in the United States. But this does not lessen the urgency of a more rapid rate of growth in the American economy. I believe that as long as we have the present immense need to raise living levels, not only for the rest of the world but also for a large segment of our own population, it is impossible for us to become overautomated. As I see it, it is not going to be easy to increase the size of our foreign aid program above the present 1 percent of national income so long as we have areas such as Harlem. The argument that aid should begin in Harlem is very hard to combat unless our economy grows at a rate at which Harlem ceases to be as much of a problem as it currently is.

Question: But does technological progress in the United States improve conditions in Harlem? I fear the contrary: we will raise the standard of living of the $20,000 a year man while conditions in Harlem may even deteriorate further.

Mr. Villard: I do not agree with you on this point, because I do not think that there is any evidence presently available of any increase in the inequality of the distribution of the national income. All the evidence that I know of points in the opposite direction. As long as this is true, increasing average income inevitably involves increasing the incomes of those in Harlem.

Question: If we do export improved technology, will it not not postpone the date when a voluntary limitation of population is achieved?

Mr. Villard: Perhaps it will. But, as I am convinced that halting population growth will take a century or so to achieve, refusing to export technology does not seem to me a practical alternative. And there is always the possibility that new goods will provide a motivation to limit the birth rate, which might not be present in a static society.

STANDARDS OF LIVING AND THE MEANING OF STATISTICS

Question: Mr. Villard talked of an income of $20,000 a year as a target, and implicitly the question has been asked: "Is that amount enough for a university professor, for example, to live on?" I would like to say that it is too much—certainly too much as a world standard. I lived on approximately a third of that as a professor in England and, I do not think that my standard of living went up one whit when I came to America, even though my income in terms of dollars increased substantially.

There are many reasons for this, and I want to emphasize that statistics tend to obscure such facts.

But my point is simply this. Would an increase in money income on the part of the entire world to the American equivalent of $20,000 really represent as much of an increase in the standard of living as such a figure seems to imply? But even more important than the question of the meaning of the statistics is the question: What do people really want? Do most people want to go in the direction of increased consumption and, if so, increased consumption of what?

Mr. Villard: I did not want to get involved in technical questions about statistics, and that is why I tried to minimize direct comparisons of per capita incomes among different countries. There is no doubt that a $50 per capita income in India is not directly comparable to an $1,800 per capita income in America; there is no doubt that the magnitude of the difference in real incomes is not as great as these figures indicate. But, on the other hand, there is no question that the difference in real income between the two countries is very great indeed.

Your other point, and its implied question, is a very different matter. You have argued that $20,000 is in some sense "too much." You have implied that many people would not like an income of that size. I can only answer that I have yet to find an American who has sufficient income. I would argue that throughout the world people want to live at least as well as people live in Harlem. In fact, however, the standard of living of people in Harlem is on the average far above that of most of the rest of the world. I gave the figure of $20,000 merely to indicate that there is still a lot of room for economic development even in America. I mentioned it in terms of the rest of the world only to establish a standard of comparison, but I do not think that *you* would like to live on less.

Comment: But on the other hand I certainly would not like to live in a world where 25 billion people spend $20,000 a year.

Mr. Villard: Nor would I—not because I don't want people to have $20,000 to spend, but because I don't want 25 billion people!

Question: The problem of comparative per capita incomes seems to me to be more complicated than this discussion implies. If the income of the average Indian, which today is $60 a year, were to go up to $90 a year in the year 2000, he would be relatively worse off in that year compared to the average American than he was in 1950, since the ratio of his income to the American's would have fallen from one thirtieth to one fiftieth. But how real, or significant, is this apparent worsening, if the $60 income is very near some lower threshold point in terms of food, shelter, health, and education, and the $90 income is significantly above that threshold point? Obviously, if a revolution of rising expectations takes place, so that people want cars and television sets, then the change from $60 to $90 will be a drop in the bucket. But if there are no such rising expectations, could not that $30 increase be significant? Moreover, even though it may not be possible to raise the overall level of income of the most underdeveloped countries by a large amount, is this true of particular items, such as food, housing, and certain types of medical care?

Mr. Villard: All I have done is to point out that, relatively, Indians may feel that their economic situation deteriorated. Let me stress again that, for the first time in human history, Marx's prediction that the rich will get richer and the poor, poorer is proving to be right. My guess is that a $90 income a year is not going to seem satisfactory to Indians when Americans are moving steadily toward $5,000.

Question: But this is what bothers me. Is this the proper criterion? I lived for a short time in a little peasant village in an Arab country. For that village these figures are meaningless. An improvement in what we might consider little things makes a crucial difference in such societies. A village of this sort does not view itself as worse off because of America's growth.

Mr. Villard: Your point might be perfectly true if decisions in India were made by peasants, but I think that they are made by people who travel. For India's leaders there has obviously been a revolution of rising expectations. If the Chinese Communists were to make significant progress in raising per capita incomes (fortunately not the case up to the present time), we might well expect serious difficulties in India. In any case, I do not think that the Indians are going to be content merely with being more adequately fed. Nor do I think that their leaders, who, after all, do know what is going on in the rest of the world, are going to be satisfied with the degree of progress India is likely to make in the next few decades.

A MORE OPTIMISTIC VIEW

Question: The situation is not quite as black as Mr. Villard's statistics would indicate. The statistics of per capita income are based on national income statistics and they leave out an element, which should be included if we are considering the future in the long run. There are many items in our national income total, which do not contribute to per capita personal incomes. I am thinking particularly of our defense expenditures. If and when these funds become available for more constructive purposes, it may become possible to relieve, to a major extent, some of the problems of poverty in Harlem and similar places.

But my grounds for optimism about the future in the long run are based upon what may be possible technologically. In the first place, we are not applying our existing technology adequately enough to the problems of the underdeveloped areas. If we were to do that, we would see a tremendous increase in income there, even without any further advances in technology. But there has been a change in the historical pattern of technology. The twentieth century has seen an institutionalization of technology. It has seen a greater emphasis on the systematic application of scientific knowledge, and, along with this, the development of large industrial research laboratories. It has seen the emergence of new sociocultural attitudes, which accept and even welcome technological change, and as a result, technology will grow at a faster pace. These developments will make long-term predictions about per capita income seem very tenuous.

Mr. Villard: Basically, you are saying that we can do better than we have done to raise per capita incomes in underdeveloped areas if we apply more of our technological knowledge to these areas. I can agree with you and at the same time ask: Why haven't we done better? We have been engaged in substantial aid programs for some time, but it is hard to find examples of governments effectively encouraging the application of advanced technology to underdeveloped areas. It is true that the Rice Institute has made progress in applying advanced technologies to the problem of growing rice, but the Institute was privately financed. On the whole, the number of similar examples is small. You say we can do more, and I agree. One of my major conclusions is that we *should* do more. But that is not altogether the issue. The issue is rather whether we will do it. And even if we do, I would still remain relatively pessimistic about the possibilities of increasing real income in underdeveloped areas in any significant way.

POPULATION CONTROL AND TECHNOLOGICAL PROGRESS

Question: Mr. Villard seems to imply that if underdeveloped countries could limit births in a similar way to that used in Ireland, there would then be a basis for economic development. Why did this not take place in Ireland?

Mr. Villard: I did not mean to imply that population limitation by itself automatically brings development. I want, rather, to emphasize that, as I see it, limitation is simply a prerequisite for any significant rise in world-wide levels of living.

Question: If we do not solve the population problem, I agree that technological progress will never enable us to support the levels of population that will ensue. But I do not think that we can say at this time whether population control will or will not be achieved. At present, it seems to me more meaningful to concentrate upon particular problems that confront the underdeveloped world. For example, if we were to concentrate, through our international agencies, upon developing a program to increase food production in different countries and also on a program to increase world trade in food, this would be a concrete step, which would be certain to aid in the solution of the population problem.

Mr. Villard: Let me emphasize again my conviction that a cessation of population growth is a prerequisite for moving towards any significant increase in world-wide living standards. That fact, alone, seems to me to justify my concentration on population problems. I am primarily interested in ultimate solutions, not in short-term palliatives.

SOCIAL AND INSTITUTIONAL BARRIERS BETWEEN
RICH AND POOR NATIONS

Question: We are faced with what amounts to two worlds—
the world of the rich, advanced Western nations and that of
the poverty-stricken, underdeveloped nations. With population,
good arable land, natural resources, and the accumulated heri-
tage of skills distributed as they are, we have the problem of
getting the wealthier nations to accept some continuing plan
for the redistribution of income. We have, in a sense, glossed
over this point by calling aid "temporary" for a whole generation now. Do social and institutional barriers prevent the
wealthier nations from facing up to the fact that they will
have to share some of their benefits with the poor nations on
a permanent basis?

Mr. Villard: Once we achieve a stable world population, we
may well consider redistributing populations as well as incomes.
But at present I think the former would be no more than a
palliative. To the question whether we are going to have to
establish aid programs on a continuing basis, I would reply:
"Yes, we *will*, but I do not know quite why we *should*." By this
I mean that we have not yet really accepted responsibility for
world-wide levels of living, any more than the Victorians accepted responsibility for providing for what they called "ne'er-
do-wells." I think our attitude toward the question of giving
international assistance is Victorian, although we have obviously
moved far from that attitude in so far as giving assistance to
our own people is concerned.

Of course, I may well be wrong as to how prepared we are
for continuing international programs—and it may also be that
poorer countries will prove to be more willing to accept continued unequal distribution of the world's income than I think.
As late as 1880, after all, 95 percent of the English people were
content to receive a mere 56 percent of the national income.
But I am convinced that the present situation is explosive, and

that it may eventually push many underdeveloped nations toward Communism. Therefore, I am wholly in favor of institutionalizing aid on a long-range basis.

Question: We have shifted the focus of our attention from primarily national considerations to international ones. As a result, we have compounded the complexity of the problems to which we addressed ourselves. I have the impression that one reason why some of us have criticized Mr. Villard's statistics and his arithmetic is to give ourselves some comfort. Perhaps the problems Mr. Villard has posed with respect to social change and technology in the international sphere cannot be solved by the present political, economic, and social order. Perhaps this structure cannot be changed until we come to the edge of an abyss.

Mr. Villard: I am not too optimistic myself. I certainly do not want to suggest that economic development is a sufficient condition for life to continue on this planet, although it is, probably, a necessary condition. On the other hand, I do not subscribe to the notion that there has to be a major and sudden change. Rather, what I think is going to happen, is a slowly developing understanding that population growth is an immense and urgent problem, that technological change is needed on a very much larger scale than in the past, and that continued international aid in various fields—including technology adapted to the special needs of underdeveloped countries—is essential. If, instead, we sit back and say that nothing can be done until we achieve social and institutional changes, then it seems to me that the changes we need will never take place.

Comment: Given the present concepts of national sovereignty, given the present world organization, and given the present social structures that are national in basis and concept, I see no way of solving the problems that you have posed except on a very fragmented and insufficient basis.

Question: Mr. Villard's analysis is focussed almost entirely on the problem of increasing income and on related questions.

When one really begins to reflect on how social change is brought about, it becomes apparent that we know very little about the processes involved. For example, we know next to nothing about how birth control is related to economic development, and next to nothing is known about the circumstances which lead people to have children. More important, I am impressed with how little we know about the extent to which, and the circumstances under which, we can, given the resources available to us, make it possible for people to move above the poverty level. Many of these questions are not primarily economic questions at all: they are questions of sociology and psychology. We know little, for example, about the expectations people have of themselves, about the extent to which they can absorb influences coming from other cultures, and so forth.

Mr. Villard: I have criticized our seminar for being provincial in concerning itself primarily with the United States. What you are doing is to criticize me in turn for being provincial in concerning myself primarily with economics!

Comment: No, I think, rather, that you are concentrating on the wrong kind of technology. You emphasize the technology of production, whereas I would emphasize the technology which brings about social change.

Mr. Villard: You are using "technology" in a sense I would not normally use. But, to answer the points you have raised: first, as far as the population question is concerned, I am in complete agreement that it is not exclusively an economic problem. We must have the help of sociologists, social psychologists, and of those who can design birth control programs for specific cultures. The ability to devise an effective birth control program for a specific culture is something we are just beginning to develop, although it is desperately needed. The world's problems obviously require a great deal more than improved technology for their solution.

Question: Part of our trouble stems from the fact that we are

trying to discuss in primarily quantitative terms the qualitative changes occurring in the world. This just simply cannot be done with any hope of success. We must give up our enchantment with the countable and begin to discuss the much more difficult and quite subjective methods needed to understand qualitative changes in ways of life—changes which are always taken for granted. If we do not succeed in dealing with qualitative changes, then the quantitative projections Mr. Villard has made will eventually become preposterously out of touch with reality.

Mr. Villard: I put forward these quantitative projections partly because I am appalled by their qualitative implications. In fact, it seems to me that ultimately the whole problem becomes qualitative. For example, how many people do we want? If all the possibilities of modern technology are put to use, New York could doubtless be made self-sufficient. But how many of us would like a world-wide population density equivalent to the present density of New York City?

Political Implications of Technology and Social Change

by JOSEPH S. CLARK
Senator from Pennsylvania

BACK IN 1956, having solved all the easy problems confronting the people of Philadelphia, I retired as mayor without seeking a second term, leaving the more difficult problems for my successor. At that time, because I had solved some of the easy problems, I was given an award known as the Philadelphia or Bok Award, which carried with it a stipend of $10,000. To my regret, the tradition had been established by earlier recipients that the winner could not spend the money on riotous living, so I was hard put to know what to do with the check. I had been thinking about some of the problems confronting the mayor, and I made a speech on the occasion of the presentation of the award on what I called "staffing freedom" [manpower utilization policy]. The rather primitive concept was that we were not putting our people in the places where they ought to go—there were too many square pegs and too many round holes and too many stock brokers and not enough Ph.D.'s.

At this point Dr. James Charlesworth of the University of Pennsylvania and of the American Academy of Political and Social Science said, "Why don't you give me the money? I will undertake to hold a series of conferences on how we can utilize manpower, and we will publish a volume of the annals of the Academy." The next thing I knew, Dr. Charlesworth held a meeting in New York at which were assembled some brains in what I had by then learned to know as the manpower utilization field.

In the meantime I had the good fortune to be elected to the Senate, and a couple of years later I got on the Committee on Labor and Public Welfare. Early in 1961, Arthur Goldberg, then the Secretary of Labor, became very much interested in the problem of manpower utilization. He said to me, "I want a bill." That gave me a bright idea, and I said to Senator Lister Hill, the Chairman of the Committee on Labor and Public Welfare, "Let's have a subcommittee and let's make me chairman." He was receptive, and so was formed the Subcommittee on Manpower and Employment.

Secretary Goldberg and I, with the able assistance of our staff, drew up the Manpower Development and Training Act that was passed in 1962; shortly thereafter Secretary Goldberg became a Supreme Court justice. Finding myself without my sponsor and with the inevitable fact that the Manpower Development and Training Act did not completely solve the problems of manpower utilization, I was persuaded by my staff, headed by my legislative assistant, Ralph Widner, to hold a series of hearings on the general subject of the effective utilization of manpower, but even more broadly than that, to explore the problems of unemployment. So we spent seven months conducting hearings—I think that it was the first comprehensive congressional investigation of the nation's manpower problems ever undertaken. We took testimony from over 150 expert witnesses, and a great number of learned papers were filed by individuals who did not choose to appear.

We investigated such subjects as the formulation of economic and manpower policy; the impact of technological change; the expansion and alterations of the labor force; the effects of discrimination and of ordinary deprivation without regard to race, creed, or color, upon the labor force; the changing demands upon the nation's educational system and—this was very important indeed—the impact of federal space and defense research upon technical and scientific manpower utilization. Finally, we investigated the question of persistent unemployment in distressed areas and the economic and employment outlook for the future.

We ended with a summary report of about 120 closely printed pages, and we also got out ten volumes, which represent the hearings and the statements of the witnesses. About half of the Subcommittee's recommendations have already been placed into effect by executive order or have been included in the 1964 and 1965 legislative recommendations of the President. From a pragmatic point of view, I think this is no mean achievement, because we really did start from scratch. As a result of the hearings, we were able to eliminate not only a substantial amount of congressional lag but a not inconsiderable amount of executive lag.

Of course we were fortunate in having Arthur Goldberg as Secretary of Labor and then William Wirtz as his successor, and, if you will excuse a partisan note, we did have in John F. Kennedy a president who had imagination and vision and a pretty good education. The momentum he created, and Arthur Goldberg kept up, and which Secretary Wirtz still keeps up, has had a very real impact.

I should observe that the remaining recommendations of the Subcommittee took a somewhat advanced political position. Neither the Congress nor the general public appear to be prepared to take action on all of the recommendations at present. One of the Subcommittee's objectives is to attempt to further educate Congress and the public concerning these needs.

Tonight I am going to confine myself to one major aspect of the Subcommittee's studies—the impact of technological change upon the work force, and, what is more important, upon public policy and pragmatic political action. We concluded that in order to minimize unemployment an exceptionally high rate of economic growth will be essential for the foreseeable future.

There are obviously two reasons why a very rapid rate of growth is essential. First, we now have what amounts to a consumer economy in which constant innovation, systematic invention, new products and markets are really essential parts of the economic process. We feel strongly that if the inventive genius and the capacity for innovation, which has characterized the American people for a couple of hundred years, should suddenly break down, we would really be in trouble. Though human wants may be virtually insatiable—and I am not at all sure that they are—the market for a particular product can become satisfied or stagnant. Modern consumer enterprises must constantly diversify and innovate in order to grow and to maintain high demand for their products. But under these conditions the economy is so vulnerable to any decrease in consumption levels, that the government must act as a sort of balancing element, keeping demand levels up with as little inflation as possible and, better still, with none.

The second reason why we are committed to a continuous high rate of economic growth is that the labor force is growing by leaps and bounds every year, so that just to retain the same level and not have the unemployment rate become really frightening, we have to create well over a million new jobs a year.

By a high rate of economic growth, I mean a rate considerably in excess of the quite exceptional rates we have been enjoying since 1961. Rapid economic growth, it is obvious, produces a number of serious problems. Workers find it difficult to adapt to changing employment opportunities. For example, a worker who joins the labor force this year will, during the course of his useful working life, probably have to change jobs at least three times. Communities are finding it difficult to cope with the changes in labor force demand already upon them.

Despite the rapid expansion of the economy in recent months, the rate of unemployment still hovers a little below 5 percent, and there are still substantial productive resources which are idle. Since both manpower and production facilities are still not totally utilized, we have experienced no appreciable inflation during the last few years.

Incidentally, I should emphasize that the 5 percent unemployment figure is quite deceptive, and lower than the actual unemployment figure, because, first of all, it ignores part-time employment, for which we do not have really effective statistical measurements and, secondly, because it does not include specific groups whose size it is difficult to judge, but who are certainly very large indeed. The figure does not include, for example, a very large number of the inhabitants of Appalachia and the upper peninsula along the Great Lakes who have stopped looking for work because they have simply become discouraged.

The question thus arises, "How fast must we grow economically to absorb the increases in labor productivity and a rapidly expanding labor force?" Over the 1909-1962 period, output per man-hour in the private sector of the economy rose an average of 2.4 percent a year. The 1947-1962 average was 3 percent. The 1958-1963 average was 3.1 percent and the average rate since 1961 has been 3.6 percent. An increase from 2 to 3 percent, perhaps 3.5 percent, in the rate of output per man-hour is more dramatic than it appears initially. With a private employment of 70 million people and a 2 percent annual productivity increase, the output of the economy must grow with sufficient rapidity to add the equivalent of 1,400,000 new jobs a year to prevent an increase in unemployment. That, to my mind, is rather frightening. And it may even be worse than that, because in the foreseeable future there will be at least 500,000, perhaps a million, new workers going into the labor force each year.

With productivity rising at the rate of 3 percent annually,

a stable output would result in the disappearance of 2,100,000 jobs. If 3.5 percent should prove to be the new trend in the long run, 2.5 million jobs would disappear each year if output were to remain stable. So the present foreseeable impact of technological change on employment, coupled with labor force increases of up to one million workers per year, clearly indicates the need for more rapid economic growth and for aggressive and imaginative efforts, public and private, to ease the adjustment process. The economy must generate from 3 million to 3.5 million new jobs a year through the rest of this decade just to keep unemployment from rising beyond its present persistent level of almost 5 percent of the work force. And frankly, gentlemen, this is an immense task.

The creation of an additional 2.5 or 3.5 million jobs would be needed to reduce unemployment to 4 percent of the labor force, and another .75 million to reach the 3 percent, which we of the Subcommittee think is an acceptable level. I am not happy with the talk we hear about 4 percent being the appropriate goal for the moment. I say that we are not doing our best unless we get the unemployment rate down to 3 percent, because we still have to take into consideration the part-time unemployed and all those people who have stopped looking for work. I think 3 percent is a maximum figure and not a minimum figure.

The figures I have just given include an allowance for workers who will re-enter the labor force as job opportunities increase, and, in my opinion, there are going to be a lot of them. The tendency is to underestimate the number of people who come back into the labor force, and the number who will move from part-time to full-time employment as job opportunities become available.

We can conclude, therefore, that the rate of economic growth necessary to create enough jobs to hold unemployment at its present level exceeds 4 percent annually. In order to reduce unemployment to 3 percent by 1968, the economy would have to expand by nearly 5.5 percent a year in constant dollars, plus whatever additional growth in money terms would be necessary to offset any increase in the price level. Now, if a level of unemployment no higher than 3 percent is to be attained by 1968 we would require a gross national product for that year of $734 billion. This means that public action must put an additional $5 billion in the pockets of consumers each year from now through 1968, or the public's own purchases of goods and services must increase by the same amount, or there must be a combination of the two which, of course, is what will happen if it is to happen at all.

With respect to this, I am sponsoring legislation that would require the President to submit to Congress each year a full employment and production budget. What does that mean? It means that the President should estimate the gross national product necessary to provide full employment and should then recommend a combination of tax and expenditure policies, in short, a fiscal policy, required to achieve that gross national product.

The Subcommittee feels that in the coming years greater emphasis will be required on the expenditure side of the budget. I regret very much, but I quite understand, the political neces- sity which caused the President to satisfy Senator Byrd and the members of the congressional establishment in charge of the appropriations committee by holding the administrative budget below the symbolic figure of $100 billion. I think he could have wisely gone at least $10 billion higher from an economic point of view, although from a political and pragmatic point of view this was probably entirely impossible.

The heavy concentration on defense and space in the fed- eral budget during the last two decades has left us with an enormous backlog of civilian public needs. These are areas where investment and new private employment could be gen- erated if activities in the public sector of the economy were funded in the federal budget. Of course, the obvious way to solve that problem (a solution which is unlikely) is to make an intelligent and sensible disarmament agreement with the Russians in order to cut back the defense budget. Until that time we are committed to throwing down the drain, from an economic point of view, those billions of dollars that otherwise could be utilized to bring our domestic economy into a position where it could be said that we really have started the Great Society on its way.

Nevertheless, these needs must be met, for they lie at the root of unemployment and of the obsolescence of public and private facilities and services in many areas of the country. By meeting these needs we would be able to provide immediate employment as well as be able to improve the economic base of local areas in order to generate long-term employment.

What I am about to say is based on the assumption, which may not be justified, that we can continue these fantastic expenditures for defense and space and still have enough left over, in terms of the productivity of the private and public sectors of the economy, to move forward on the domestic front and take the actions necessary to bring unemployment to an acceptable level. There are many areas that can absorb economic resources, which are right there to be used, if the fiscal problem can be solved. The allocation of resources to community development, housing for example, has never been governed by considerations of how it might affect annual unemployment levels. Yet programs of this type, having a very high labor content per dollar spent, are an important part of the job-creating impetus that federal expenditures can give. Witnesses before the Subcommittee estimated that the labor content of urban housing is somewhere between $560 to $660 per thousand dollars of contract award. This is the area where, in my opinion, the unskilled portion of the labor force can most effectively be either re-employed or employed in the first instance.

The field of community development in general, which includes housing, urban renewal, mass transit, highways, air and water pollution control, sewage systems, sewage disposal plants, and other local public facilities, represents one of the largest areas of the unmet needs of the nation. There are presently between seven and eight million substandard dwelling units in our urban areas. The war baby generation has now reached maturity and will soon be marrying and starting to raise families. The demand for new and rehabilitated dwellings, therefore, will increase dramatically. In addition, the growing proportion of older citizens in the population calls for the construction of housing especially adapted to their needs—such housing is not currently available.

An additional need for more low rent or low cost housing units is evidenced by the number of families who are unwillingly doubling up for economic reasons. I think the need for additional public housing is almost unmeasurable so long as one fifth of the nation remains ill-housed, ill-fed, and ill-clad.

There are, in addition, enormous numbers of obsolete commercial and industrial structures in our metropolitan areas. Continuous expansion of the metropolitan areas will increasingly direct attention toward improving inadequate transportation facilities, whether bus, subway, or railroad commuter services, as well as toward stepping up construction of roads, bridges, and tunnels. Of the future success of this attempt to accomodate mushrooming automobile traffic, I take a very dim view.

Additional investment to support metropolitan growth is essential. Schools, colleges, hospitals, health services, playgrounds, recreation buildings, welfare institutions, and the like will be needed. And yet in the housing bill for 1964 the President, crippled by the $100 billion budget ceiling, recommended only $100 million for community facilities. Some of us in the Senate offered an amendment that raised the $100 million to $2 billion, which is a small part of what we need. I invite your speculation, in the light of the present economic views of Congress, on our chances of getting this amendment passed. They are almost nil.

And yet the $900 million accelerated public works program President Kennedy sponsored several years ago was a fantastic success. I could cite you page, chapter, and verse in Pennsylvania, where it resulted in the construction of wealth producing facilities, in sanitation facilities that were badly needed, and in health facilities, and at the same time made a very substantial contribution to the reduction of unemployment. To my chagrin that program was allowed to lapse. In 1965 it was given a very modest new lease on life. I think, perhaps, the number one priority is to get back to a massive program of public works in which we think not only of lifting the face of the country, but also of putting people to work.

The sums that will be required just to keep up with the population growth in urban areas are really staggering. Estimates indicate than an investment of between $500 billion and $700 billion (both from private and public sources) will be necessary just to accomodate the increasing urban population over the next twenty years. And if the elimination of existing blight and obsolescence is included, an average annual investment of from $110 billion to $125 billion in private and public funds will be required each year over the next two decades in order to preserve and restore the livability of the metropolitan areas of the country.

Parenthetically I should note that in 1961 the nation spent a total of about $50 billion, including both private and public expenditures, on investment. The sum is more now but not by very much. And the sums I have been talking about are just meant to meet the needs of the urban areas. In the sparsely populated areas of the country urgent needs also exist. One study presented to our Subcommittee, gives $4 billion per annum as the total amount of federal investment required over a ten-year period for the development of rivers and harbors, for forest land improvements, for soil conservation, for recreation facilities, and for other resource developments. That is about twice the current rate of expenditure. About 250,000 new jobs would be generated by the additional $2 billion allocated over a decade for just this purpose. Of course that is chicken feed in terms of what we have been talking about. But several years ago a study, carried out under the auspices of Resources for the Future, Inc., a research project sponsored by the Ford Foundation, estimated that the backlog of needed work programs in forestry, soil and water conservation, preservation of fish and wildlife, range revegetation, and other similar investments would require about $20 billion. This calculation did not include the heavy construction programs in water development, such as dams and the like, and I should note that a backlog of authorized projects, requiring nearly as great a sum, has now been cleared by the Corps of Engineers and is stacked up in the Public Works Committees of the Congresses; these projects cannot be started because of fiscal limitations.

Water and soil resource development needs are vast. Total requirements for withdrawal of water from lakes and streams in the United States in 1954 was 300 billion gallons daily, or only 27 percent of the stream flow. Withdrawals in 1980 will be, it is estimated, 559 billion gallons daily, and by the year 2000 more than 888 billion gallons daily, or more than 80 percent of the total normal stream flow. Water programs up to 1980 total almost $55 billion for federal programs and $173 billion for nonfederal programs.

Some of the programs for the expansion and development of water facilities are based on the urgent need to overcome accumulated backlogs. For example, the Public Health Service estimates that to overcome the backlog of municipal sewage treatment facilities by 1971 would require $600 million annually, or more than twice the present rate of expenditure. I can testify personally to the really pathetic condition of the sewage disposal treatment facilities and sewage disposal treatment plants in the Commonwealth of Pennsylvania, due in large part to bankruptcy of local municipalities and to the inadequacy of the state and the federal contributions. All three of our major river basins, the Delaware, the Susquehanna, and the Ohio are so polluted that it is difficult indeed for even the hardiest catfish to exist.

The Subcommittee was able to discover numerous highly technical fields in which the expertise gained in the defense and space efforts might be applied, for instance, in world-wide, detailed inspection and surveillance systems for arms control, in new educational technology for classrooms, in new aids for air traffic control, in medical electronics, in development of new energy sources, in mass transportation, and in one which is a favorite of mine, oceanography.

We have a national obligation, I think, to bring the many lessons we have learned from defense industries to bear upon the continuing unmet needs of our communities and people; needs, which the defense industries with their advanced technologies may be able to help meet in this century of technological revolution.

We have certainly seen in the last decade, in defense and in space activities, the most remarkable achievement of national goals, hardly dreamt of in 1955, unheard of in 1945. We have assembled expert teams of engineers, research men, industrialists, and production forces to meet the new national goals in space, communication, and national security. I believe that the time is soon approaching when we must harness this energy and technical skill to improve the conditions of human life. There are millions of Americans who wish simply to live in a safe and sanitary house they can afford to buy or rent. Our technology can help them do that. Millions of Americans would benefit from urban and interurban mass transportation systems.

I believe that the experts of a nation, who can put tons of instruments on the moon and send a man around the earth in ninety minutes, ought to be able to get commuters comfortably to their jobs thirty miles away and back home again in one half or one quarter of the time presently spent by millions of citizens driving bumper to bumper on our most modern and expensive freeways. Millions of American children are waiting for twentieth century technology to provide them with adequate schools, educational facilities, and teaching aids. It seems to me that a communications system that can flash a television picture around the earth ought to be able to solve the communication problems in our schools and provide topnotch information and educational service to millions of school children.

These are some of the questions which challenge our economy. The new jobs we need so desperately will not come from the application of sophisticated technology to traditional productive enterprise, but they could come from the application of technology to a whole new area of endeavor and unmet need.

The capabilities of our technological society are so great, and its real problems are so incredibly complex that we cannot hope to utilize its potential without some major departures from the usual in both public and private policies. I suggest that we apply the principle of the systems approach that has been so successful in defense and in space activities to the solution of the more pressing human problems in the United States. Such an undertaking would mean that we should have to engage in some of the long-range national planning and coordination that most of the nations of Eastern Europe have adopted, some before and all of them since the end of World War II.

I suggest that an essential defect of America's national economic program is inadequate national planning. There must be some direction, some coordination. Somebody must set national objectives and performance levels. And nowhere is this more necessary than in the attempt to bring our technological capabilities to bear on some of our most chronic public needs. Change is the only permanent part of the modern scene. In a static society anticipation and foresight are unnecessary, but when the daily lives of 200 million people, and it will soon be that, are in constant flux, when their individual welfare is threatened by forces beyond their control, then planning and definition of social objectives become essential. And only government can provide that perspective for the community as a whole without regard to personal vested interests.

John Dewey once wrote that a culture which permits science to destroy its values without permitting science to create new ones is a culture which destroys itself. Now, in the early days of the second industrial revolution—the manpower revolution as I call it—we have an opportunity to create these new political values, which will carry us towards our objectives. We have the promise, in the words of W. W. Rostow, of seeing what man can and will do when the burden of scarcity is lifted from him.

We built this nation, said the President in his 1964 State of the Union message, to serve its people. We want to grow, and build, and create. But we want progress to be the servant and not the master of men. We do not intend to live in the midst of abundance, isolated from neighbors and nature, confined to blighted cities and bleak suburbs, stunted by a poverty of learning and an emptiness of leisure. The Great Society asks not only how to create wealth, but how to use it; not only how fast we are going, but where we are headed. The most dangerous phenomenon in American life today is a social and political lag, the lag between what our society requires and what our political institutions are prepared to provide.

Our purpose must be to build a world in which technological innovation and efficiency are not ends in themselves. Our purpose should be the achievement of freedom, intellectual excellence, and the perfection of the community. The test of government of, by, and for the people is how well it will respond to this challenge. So I end where I started. Those of us who are interested in the manpower revolution are interested in the proper utilization of American manpower and woman-power.

DISCUSSION

THE POLITICAL STAGE: LEGISLATIVE ACTION, CONGRESSIONAL REFORM

Question: Could you say something more about the national manpower and full-employment budget?

Senator Clark: The manpower and full-employment budget could well accompany the administrative budget and the report of the Council of Economic Advisors. It would lay stress on the utilization of manpower and how to deal with the problem of full employment. It would involve, I think, some fairly important amendments to the Employment Act of 1946 and it would actually call for another major document by the president at the start of each congressional year. In the best of all worlds, the manpower and full-employment budget would become an integral part of the economic report of the president and would fuse with the budget message. I would hope that in the fore-seeable future, we will be able to get away from the administrative budget, which is really deceptive and hampering.

No well-run city in the country, no well-run business for that matter, would permit a budget similar to our federal administrative budget. They would snarl at its primitiveness and stupidity. The federal administrative budget, however, is part of a myth, part of the symbolism of the nation's economic life. And such is the mythology of the country, that when the administrative budget shows a deficit a perceptible number of the voting public whose support is essential to political success takes fright.

Once it becomes a legislative requirement that we state our manpower utilization budget in terms of how large a gross national product is necessary to secure full employment, and that we give appropriate consideration to the threat of inflation, then we can get a presentation to the Congress and to the people of an overall national plan for each year which has as its objective the achievement of maximum production, maximum employment, and reasonable price stability. The emphasis, however, would be on people and employment, rather than on figures and deficits.

Comment: I hope that Senator Clark realizes how useful and influential the volumes coming out of the manpower hearings have been. They constitute an enormously useful literature, which will have an influence for a number of years to come. If the same thing can be done in the area of the full-employment budget the influence will go far beyond the halls of Congress.

Question: While we are discussing the full-employment budget I want to bring up the subject of indicative planning. It seems to me that President Johnson has already been engaged in indicative planning. He has gone to the major businesses and set goals for Negro employment. Much the same thing was done with regard to the balance of payments problem, and it occurs to me that the president may have carried the technique over from one place to the other. It seems to be a way of working that is congenial to him and might be the basis on which some progress on manpower utilization might take place.

Senator Clark: It may very well be, and I certainly hope so, but I must add that I myself am skeptical about the effects of exhortation, particularly upon the business community. I hope that it does not end in a confrontation similar to that between President Kennedy and Mr. Blough over steel prices.

I should add that hundreds of businessmen these days, many of them graduates of the better business schools of the country, have been taught other things than how to make the most rapid profit. It may very well be that we have arrived at the age of business enlightenment in which this sort of exhortation will be effective.

Question: You have established the fact that the faster we run economically, the further we have to look ahead. We have asked in the course of discussions during the seminar how to get people to engage in programs, particularly in the political arena, that promise to pay off beyond their own tenure of office. What are the political mechanisms that will enable us to plan for long-range success, so that we do not get too involved in short-range success, and long-range suicide?

Senator Clark: I would say an imperative prerequisite for all the things we have discussed is congressional reform. By and large the executive and the judicial arms of our government have done a pretty good job over the years, but the Congress has almost always done a quite unsatisfactory job. This was as true in the days of George Washington as it is now. But we could afford it in Washington's day, and we simply cannot afford it today. The world has shrunk, the process of change has speeded up, and we can no longer afford the built-in inhibitions, the red tape, the rules, the procedures, the practices, the traditions, and, indeed, the criteria by which congressmen are selected and serve and continue in office.

Question: Several years ago I heard you and Dr. Gunnar Myrdal speak at a convocation of the Fund for the Republic. At that time Dr. Myrdal took a very similar position to the one you have taken tonight, not only on the need for more direct involvement by the government in planning for full employment and production, but also on the need for something approximating the type of national indicative planning that Western Europe and now to some degree Canada is employing. I recall that you not only seconded Dr. Myrdal's proposals, but that you also called attention to the fact that this would be very difficult to accomplish under our present form of government. I wonder if you might expand somewhat on what kinds of changes in our political system you think might be necessary to accomplish these goals.

Senator Clark: You may know that we brought over from Sweden three or four of their top economists, not only in government but also in management and labor, who were a great help to us. But to answer your questions more directly—I sometimes despair of making any really significant changes in our form of government. I cannot see that in my lifetime, or, indeed, in yours, we will be able to get rid of what I consider the utterly impossible separation of powers theory which Montesquieu and the Founding Fathers saddled upon us. I do think, however, that over a period of years, merely by plugging away with the long, slow educational process, the solutions that I have suggested tonight will almost inevitably come to fruition, even if we have to maintain very large defense and space expenditures. If we can come to a meaningful arms control agreement—and I am much more optimistic about this than most people—then it will be possible both to cut taxes and to carry a number of these programs through.

I should add that at the present time the control of our subcommittee on labor is in the hands of a well-informed group of liberals. We are in a position to bring to the floor of the Senate almost any bill that a concensus of economists will support. And once we get a bill on the floor, we have a very good chance of passing it.

Question: It seems to me that full-employment budget legislation is very close in character to the original Murray Bill, before it was amended and then became the Employment Act of 1946. It may be that the debates of that time furnish a dress rehearsal of what the arguments would be with respect to the bill that you are introducing.

Senator Clark: Yes, except for this crucial difference: There is a far greater necessity for it now.

Question: One of the reforms needed in the federal budget

area, in addition to the manpower budget, is a budget that would distinguish between those items of governmental expense that occur in day to day operations and those items that involve investment projects.

Senator Clark: This distinction is already being made in our present budget, but, unfortunately, it is buried in the middle of the budget so that reporters do not find it and may not understand the significance of the distinction when they do discover it. The only question reporters tend to ask is what, in conventional terms, is the deficit or the surplus according to the accounting monstrosity called the administratve budget.

TRAINING PROGRAMS UNDER THE MANPOWER DEVELOPMENT
AND TRAINING ACT

Question: What is your assessment of the retraining programs being carried out under the provisions of the Manpower Development and Training Act?

Senator Clark: I am disappointed with the results shown by the training programs up to this time, but I am not entirely sure that my dissappointment is justified. What disappoints me most is the almost total failure of the states to pick up the program, to contribute the matching funds, and to move zealously ahead; a failure which is particularly pronounced in those states where there is chronic unemployment. This means that we will have what to all intents and purposes is a 100 percent federally financed program. Actually, the matching funds are relatively minimal, because the program in terms of dollars involved is not very large.

Another element that disturbs me is our inability to find an administrative solution. To carry out the program successfully, we will need to centralize responsibility and authority. Look at the situation confronting the administrators of the program now. In the first place, either the Department of Labor or the Area Redevelopment Administration working out of the Department of Commerce, has to ascertain that there are a specific number of trainable unemployed in a particular city or district or area suffering from chronic and persistent unemployment. Then it must be ascertained that retraining is practical and potential job opportunities have to be found for them. After that, the particular program having been developed, it comes under the administration of the Department of Health, Education, and Welfare. From there the program moves to the local school districts through the state's department of education. Then the administrators have to go to the Bureau of Employment Security of the state, which is financed by federal funds, but staffed with personnel responsible to the state involved. Because the administrative structure is so complex and diffused, it does not surprise me that the program has not been more successful. And I should add, incidentally, that a large number of people have been successfully retrained. And of those people who complete a course of retraining, Secretary Wirtz says that approximately 70 percent obtain employment.

I have up to this point told you some of my disappointments. I should in fairness to the program emphasize its promise and what it can achieve in particular cases. I have seen one of the schools in Pittsburgh where they were training people in electronics. These were young people who came from tragic communities like Briarcliff in Fayette County, Pennsylvania. Briarcliff is a village perched on top of an abandoned coal mine. Most of the eleven hundred or so people in the community are on relief. The men have been to Cleveland, Detroit, Pittsburgh, and other cities, looking for jobs, but the great majority cannot find any because they have no education and only one skill, that of the coal miner. Some of the young men from Briarcliff, who are able to live on the training allowance, can come to Pittsburgh and take these electronic courses. This is a sight to make ones heart leap—it is such an obviously good thing.

Question: You have said you are impressed by the fact that when it comes to the problem of space exploration we can mobilize personnel, imagination, creativity, and so forth to achieve our goal, but that when it comes to human problems this mobilization does not seem possible. I would suggest that one hypothesis might be that when we deal with space problems we are, basically, operating on a national level; when we are dealing with human problems we are, in fact, dealing with problems on a local or state level, a level you have so aptly described.

If we start a manpower training program, to what extent would it be possible to take part of the program and try to cut through these local problems to get some kind of regional political cooperation? Otherwise we may be simply pouring billions of dollars down the drain replicating mediocre vocational educational programs in small communities.

Senator Clark: I have some comments to make about this problem. Social scientists and historians will recall that in the late nineteenth century Lord Bryce said that municipal government was the great failure in America. I have thought for some time that another great failure of American government was state government.

I can give you discouraging examples of the problems encountered. Allegheny County in Pennsylvania is one of the most disheartening. There are 283 units of local government in Allegheny County. About fifty years ago, a dedicated public servant conceived the thoroughly sound plan of consolidating them into the city of Pittsburgh. Allegheny County and perhaps three or four other *ad hoc* agencies were to deal with particular problems such as mass transit and the like. After a great fanfare the consolidation proposal was put up to the voters of Allegheny County who rejected it by a vote of more than two to one. You can be sure that all the vested political interests in each of the 283 units of local government were out getting all their friends to vote no; you can be sure that those living in the snug harbor of the well-to-do suburbs, who did not want to be merged with the minority groups concentrated in the city of Pittsburgh, voted no.

However, sometimes it is possible to bring about certain kinds of cooperation. Those of you who are interested in the increasingly acute problem of water supply will recall that finally, just in the nick of time, we were able to create a commission consisting of representatives from the four states involved to control the use of every drop of water coming into the Delaware Basin. We are working on the same thing for the Susquehanna, and next in line will be the area of western Pennsylvania, where we will get to work on the Allegheny and the Monongohela. These *ad hoc* solutions are, in my opinion, the best hope, even though they are fairly primitive and not too satisfactory in many cases. The only metropolitan area, as far as I know, which has had any success at all in creating a unit of government large enough to be able to deal with the problems we have been discussing tonight is the city of Atlanta, where two very brilliant mayors, during a period of sixteen years, did get practically all of Fulton County with the limits of their city.

There are some provisions in the Housing Act of 1964, which would vastly encourage cooperative metropolitan area planning and particularly cooperative land use. On the state level, we now have legislation in Pennsylvania, which encourages the creation of county planning boards with some paid staff and with the potential of creating county zoning ordinance standards. For example, in southeastern Pennsylvania the counties of Montgomery, Bucks, Delaware, and Chester have combined for this purpose with Philadelphia County. Each county has a zoning code and each is working on a master plan for land use, prepared by a coordinated group of planners from each of the five units of local government. But the problem is that the city of Philadelphia is largely populated by minority groups and the surrounding counties are all strongly Republican and dominated by relatively well-to-do white Protestants who tend to loathe the city of Philadelphia. The problem of securing cooperation in such a case is very difficult indeed. We were able to get cooperation to some extent and, in my opinion we did better than any other metropolitan area in the country, as far as mass transit is concerned. This was because Philadelphia County was able to get Bucks County and Montgomery County to agree to a tri-county plan under which the suburban lines of the Pennsylvania and Reading Railroad were leased to a nonprofit corporation run by the three counties. The result has been a vast reduction in fares, the purchase of a number of new and very comfortable aluminum commuting cars, and an extraordinary movement from the highways to this suburban mass transit system by the people who live in these counties. Recently Delaware County also joined the system.

THE EFFECTIVE UTILIZATION OF THE HIGHLY SKILLED AND
THE HIGHLY EDUCATED

Question: It seems to me that staffing is the really difficult problem and in a much larger sense than Senator Clark has in mind. The problem is finding niches for people with significant ideas so that they can transform ideas into action. People who have the ability, the interest, and the right position will be able to come up with all sorts of ideas. Legislators will not have to tell them, "You ought to use developments derived from space technology for this, and from atomic energy for that." They will work out these things for themselves. What I have missed in the Senator's remarks is how to get the individuals with ideas into the right positions to get us moving.

Senator Clark: The picture of manpower I have is that of a pyramid; at the bottom are the unskilled and at the top are the highly skilled. The real job of staffing, as I originally conceived it, was to make sure that we were training skilled minds in the areas which were in potential short supply, in the light of the foreseeable needs of our society. I conceived a program based on work done by Ewan Clague in the Bureau of Census on the probable future need for skilled manpower. This need would be estimated, to the extent that it is possible, for five years ahead, hopefully for ten years. However, our society is so dynamic that it is never possible to know when it will move off on another line of development. These forecasts were to be made available to guidance counselors in schools and colleges, so that when young people were attempting to determine what career they should follow, a strong argument in favor of those fields in which they could hope to use their talents most fruitfully would be presented to them.

This is what I really mean by manpower planning. To give you one example of what I have in mind, the generation to come will need an almost indefinite number of oceanographers, and yet very few universities are teaching oceanography.

Comment: But a footnote should be added. It is easier to train oceanographers than it is to train people who know that we need oceanographers or know how best to use the oceanographers we do have.

GOVERNMENTAL REFORM: DEMOCRACY AND THE CREATION OF ELITE DECISION-MAKERS

Question: The thrust of Senator Clark's excellent book,[1] on the reform of Congress, is the need for more experts to reduce bureaucratic and congressional mediocrity. What concerns me is his suggestion that the federal government, Congress and especially, the executive branch, become increasingly dominated by an elite group, immune from the popular distortion of the mass media and from the rather blunting influence of a

[1] Congress: The Sapless Branch (New York, Harper, 1964); revised ed., paper (New York, Harper, 1965).

not-too-well informed public opinion. This has happened in the area of national defense. One finds that members of Congress on the Armed Services Committee, on the Space and Aeronautics Committee, and, most strikingly, on the Joint Committee on Atomic Energy have become technocrats. But I want to emphasize that we have lost something of democratic decision-making in the process.

Senator Clark: I think that a lot of what has been accomplished in the Defense Department has been done in spite of, instead of with the help of, Congress. And I do not think that the Armed Services Committee has always been too sympathetic with what has taken place in the Defense Department. Let me give examples drawn from my own experience. The big practical political problem for me, raised by cost effectiveness as it is applied by the Defense Department, is what installations in Pennsylvania Secretary MacNamara is going to close in the interest of efficiency and economy in government. Two were recently threatened, one being the Philadelphia Navy Yard, the other the Olmstead Air Force Base outside Harrisburg. A navy yard, perhaps two, obviously had to be closed somewhere in the United States. I thought the advantages of the Philadelphia Navy Yard were such that it was wiser not to close it, but obviously I was biased.

But let us look at my bias. If the Philadelphia Navy Yard had been closed, 7000 jobs would have been lost in that area, and who would have been blamed? Not Secretary MacNamara but Senator Clark. The voters would have said, "Look, they did not close the Brooklyn Navy Yard and why is that? It is because Senator Robert Kennedy is a more powerful and important figure in Washington than Senator Joseph Clark." As a matter of fact, the decision had nothing to do with the relative power of the two senators—actually, it was made on the merits of the case. But as a practical matter of everyday political life I am always forced, if I want to stay in office, to try to persuade the Department of Defense to keep installations in Pennsylvania in operation. I think you can appreciate that the congressional animal is a very different creature under our form of government than the executive animal. A man with courage and drive like Secretary MacNamara can do what he does, partly because he has only one superior, the President of the United States, who is committed to MacNamara's program and will back him up.

Question: Senator Clark has pointed out that we could obliterate slums, but that we are not willing to pay what it would cost. He has suggested that we need something like a systems approach when we are tackling large problems. I have heard a number of people appeal for a systems approach to be applied to human relations and social studies. But no one indicates exactly what kind of systems approach is to be utilized in these areas. We know the effect and value of a systems approach in something like aerospace engineering but it seems that we are much less clear about what the term means when it is applied to human engineering problems.

Senator Clark: James Webb, the administrator of National Aeronautics and Space Administration, has said to me, "We can solve the unemployment problem in Pennsylvania. By utilizing the systems techniques, which we apply in the space industry, we can rehabilitate all the economically depressed areas in your state."

I would suggest that as a start some of you sitting around this table might indicate a few of the details. I am sure that there are other groups and individuals who have already begun to block in the outlines of a systems approach to major social problems.

THE QUESTION OF GROWTH RATES: ECONOMY AND POPULATION

Question: People talk a good deal of a 2 percent rate of population growth per annum and a 4 percent rate of economic growth. If these growth rates are extrapolated for a hundred years, which even by American standards is not an eternity, we end up with extraordinarily large increases. Over somewhat longer periods of time, the figures for increases in wealth and output become a hundred or a thousand times as large as our present wealth and output. How do you envisage a gross national product that is a hundred or a thousand times larger than the present?

Senator Clark: I do not envisage such a gross national product. All I have done is to give you the projections for the immediate future furnished by some quite able people. But I would suggest that you take a look at Tennyson's "Locksley Hall," in which the poet talks about the airplane, the possibility of going to the moon, and predicts an enormous growth in international trade. In 1842, when these predictions were made, they must have seemed starry-eyed. I do not claim to have the imagination of a poet. All I can do is to take a very cautious peek at 1970.

Question: But some of these long-term developments are not as far ahead in the future as you imply. For example, I have read an early draft of the Full-Employment Bill that you are sponsoring, and it seems to me that you are still talking about maximum employment. I wonder why you did not choose to introduce another facet of the problem and advocate that one of the tasks of a presidential report on manpower should be to recommend how many people could be taken out of the labor force because they were no longer needed for production and could, therefore, be made available for other activities. They could, for example, pursue their education further. This question is also related to the problem of changing the pattern of the distribution of income. It also relates to the problem of the returning Peace Corps volunteers—many of them would go on doing voluntary work if they could.

Senator Clark: It seems to me that your proposal, interesting as it may be, is so far out of the realm of political pragmatism at the present time that there is simply no chance of getting anything, which accepts this line of approach, into the bill.

Question: But let us pursue it here, and take it one step further. The Secretary of Labor has indicated to me that in his opinion the real progress in our approach to manpower problems may be occurring in education. We are near the point where we can talk about a man in school or college as being at work just in the same way as about somebody at work at a conventional job.

Senator Clark: I think that this is a concept that is so far removed from the mores of Americans that the answer you would get to your suggestion is, "Let's not start paying people to stay in college until we have enough teachers to teach them."

SPECIFIC AREAS OF POTENTIAL EMPLOYMENT: CONSTRUCTION

Question: Senator Clark has mentioned that construction activity is an area where a good deal of employment can be created and has called attention to the fact that there is a high ratio in that area at the present time between dollars expended and wages paid. I am all for getting more and more people into the field of construction, if this is a field where large employment is possible. And I am all for getting more and more very bright people into this area in order to bring about necessary technological developments. But I would emphasize that one of the obvious results of putting very bright people into this area is that construction will become automated at long last, and, as a result, construction will become a segment of our economy where employment opportunities may decrease rather than increase.

Senator Clark: You may be very right, in the long-run. But let us take the intermediate gain. I hope that automation or cybernation or what have you will eventually totally replace the unskilled laborer. It has made great inroads in construction already. We used to dig ditches with immigrant labor and now we dig them with bulldozers. Still, as you look over the economy for areas which offer potential employment for the unskilled worker, I have a strong impression that the construction field offers the greatest opportunity for putting large numbers of unskilled people to work. This is a problem that this generation will have to tackle in the next week or month or year, and I believe we should take advantage of the opportunity while we have it.

Question: In construction we are still in the medieval stage of production, and I do not believe that this situation can last much longer.

Senator Clark: The outcome depends a great deal on the strength of the building trades unions. This is a good example of a point I made previously; you need some very able people at the top of the pyramid, in this case in construction man-power, in order to direct activities intelligently, just as you do in the railroad industry (and, parenthetically, it is my impression that both industries suffer from a shortage of such able people), but in addition to these able people potential employment for a great number of people who are not so able or so highly skilled still exists.

THE SPACE ENDEAVOR AS THE MORAL AND TECHNOLOGICAL
EQUIVALENT OF WAR

Question: Some of the comments made tonight seem to imply that we ought to cut back our space activities in order to release resources for more immediate and tangible social goals. I suggest that we should not fight the space endeavor. This activity may not only be a moral equivalent of war, but it also may be a technological equivalent of war because it leads to techno-logical developments in much the same way that military activities have and still do. Look, for example, at what space activity has done to raise our educational goals and achievements since 1957.

Senator Clark: I do not disagree with your general point, but I would like to make a tangential comment on it. When a ceiling is fixed at 100 billion dollars in the administrative budget, you find yourself making very difficult choices. How much of the 100 billion dollars should go to space, how much to armaments, how much to education, and how much to highways. My plea would be, do not destroy the space program, but do give education a lot more money than it is currently getting.

I think the critical problem in the space endeavor is one of pace. Would we not have a much more effective space program if we spaced it a little further out instead of treating it in terms of a crash program whose principal motivation is national prestige?

THE UTILIZATION OF OBSOLETE GOVERNMENT FACILITIES

Question: Could any method be developed to direct our obsolete or redundant defense establishments to some other type of needed public activity? Could the Secretary of Defense be made responsible to check thoroughly for projects that could be established in facilities like these?

Senator Clark: This is being done today, and in my opinion being done quite well. When the Olmstead Air Force Base was closed, a bureau in the Defense Department made studies and came up with a definite promise that every worker at the Olmstead Air Force Base would be given a job at the same wages elsewhere.

Question: But such a procedure means breaking up the organization.

Senator Clark: The maintenance of the organization as a going concern presents a very difficult problem indeed, even without the cut in the defense budget which might come from any plan leading to arms control. The best brains in the Defense Department, not to mention in the manpower field in general, have been devoted to finding ways and means of working out solutions for this kind of problem. Obviously a massive public works program is one of the best ways of meeting it.

LIMITS TO THE APPLICATION OF TECHNOLOGY

Question: Senator Clark has asked, "Why can't we learn how to put advanced technologies into operation? For example, why don't we have advanced educational audio-visual aids in every classroom in the country, since we have the technological capacity to transmit television pictures around the world by satellite?" The point is, technology can already do just this. The limiting factor is not technology; it is our lack of imagination and our indecisiveness in allocating our resources.

Senator Clark: I thoroughly agree with you. It seems to me that what is needed now is not technology, but, rather, technological engineers and salesmen who will bring across, for example, to the Lancaster County school board in Pennsylvania, an understanding of the potentialities of technology of which they have never yet dreamed.

Participants in the Seminar, 1964-1965

R. CHRISTIAN ANDERSON
Brookhaven National Laboratory

STEFAN BAUER-MENGELBERG
International Business Machines
  Corporation

ARNOLD BEICHMAN
Electrical Union World

DANIEL BELL
Department of Sociology
Columbia University

MYRON BLOY, JR.
The Protestant Ministry
Massachusetts Institute of Technology

CHARLES R. BOWEN
International Business Machines
  Corporation

PETER B. BUCK
National Academy of Sciences

A. C. BURSTEIN
Department of Commerce
City of New York

NEIL W. CHAMBERLAIN
Department of Economics
Yale University

EDWARD T. CHASE
New American Library

EWAN CLAGUE
Bureau of Labor Statistics

THOMAS E. COONEY, JR.
Science and Engineering Division
Ford Foundation

CHARLES DeCARLO
International Business Machines
  Corporation

FREDERIC de HOFFMANN
General Dynamics Corporation

SAMUEL DEVONS
Department of Physics
Columbia University

ALFRED S. EICHNER
Department of Economics
Columbia University

LUTHER H. EVANS
International and Legal Collections
Columbia University

VICTOR R. FUCHS
National Bureau of Economic
  Research

EDWIN GEE
E. I. du Pont de Nemours and
  Company

ELI GINZBERG
Graduate School of Business
Columbia University

WALTER GOLDSTEIN
Department of Political Science
Brooklyn College

GEORGE A. GRAHAM
Brookings Institution

DANIEL GREENBERG
Department of History
Columbia University

ROBERT H. GUEST
Tufts School of Business Administration, Dartmouth College

DALE HIESTAND
Graduate School of Business
Columbia University

MARY ALICE HILTON
Institute for Cybercultural Research

RONALD E. JABLONSKI
School of Business
University of Michigan

EARL D. JOHNSON
Consultant

DAVID KAPLAN
Economics of Distribution Foundation

NORMAN KAPLAN
Department of Sociology
University of Pennsylvania

JACOB J. KAUFMAN
Department of Economics
Pennsylvania State University

MELVIN KRANZBERG
Department of History
Case Institute of Technology

JAMES W. KUHN
Graduate School of Business
Columbia University

NORMAN KURLAND
New York State
  Department of Education

HARVEY J. LEVIN
Department of Economics
Hofstra University

LEWIS LORWIN
Consultant

JOHN McCOLLUM
U.S. Department of Health,
  Education, and Welfare

SILVIA McCOLLUM
Bureau of Labor Statistics
U.S. Department of Labor

EMMANUEL MESTHENE
University Program on Technology
  and Society
Harvard University

DONALD N. MICHAEL
Institute for Policy Studies

CHARLOTTE MULLER
School of Public Health and
  Administrative Medicine
Columbia University

PAUL NORGREEN
Department of Economics
Rutgers University

VICTOR PASCHKIS
School of Engineering
Columbia University

GEORGE W. PETRIE
International Business Machines
  Corporation

E. R. PIORE
International Business Machines
  Corporation

GIULIO PONTECORVO
Graduate School of Business
Columbia University

I. I. RABI
University Professor
Columbia University

GEORGE REYNOLDS
Department of Physics
Princeton University

ORMSBEE W. ROBINSON
International Business Machines
  Corporation

VIRGIL ROGERS
National Education Association

DEAN H. ROSENSTEEL
American Management Association

MARIO G. SALVADORI
School of Engineering
Columbia University

DAVID SIDORSKY
Department of Philosophy
Columbia University

HAROLD J. SZOLD
Lehman Brothers

FRANK TANNENBAUM
University Seminars
Columbia University

ROBERT G. THEOBALD
Consultant

MARKO TURITZ
California-Texas Oil Corporation

ALFRED VAN TASSEL
Department of Economics
Columbia University

HENRY H. VILLARD
Department of Economics
The City College of New York

AARON W. WARNER
Department of Economics
Columbia University

E. KIRBY WARREN
Graduate School of Business
Columbia University

SEYMOUR WOLFBEIN
U.S. Department of Labor

CHRISTOPHER WRIGHT
Council for Atomic Age Studies
Columbia University